设计学系列成果专著

任文东　主编

跨学科路径下智能服装设计与教育策略研究

DESIGN PLUS X:
AN INTERDISCIPLINARY
APPROACH TO
EDUCATION STRATEGY
IN FASHION DESIGN
WITH SPECIFIC FOCUS
ON SMART CLOTHING

丁　玮　著

中国纺织出版社有限公司

总序
FOREWORD

当今时代是全球科学技术、文化艺术快速发展的重要历史时期，也是艺术设计发展取得突破性成就的黄金时代。随着计算机信息技术的迅猛发展，人类社会逐步开启了全新的世界观及生活观，前沿科技彻底颠覆了工业社会时代设计哲学指导的设计范围、设计内容、设计意义。当今设计所面临的是一个多元交叉、领域交融、机遇与挑战并存的新时代，探索设计与设计教育的新理念、研究未来设计学科发展的新范式在当下具有非常重要和切实的意义。

一个新学科的兴起预示着更多学科的交叉与融合。这种融合不仅发生在不同国家不同文化上，还发生在新的技术与科学的加入上，所以多元化学科交叉与融合将是艺术设计未来的发展趋势。任何学科都需要有创新力，设计更是如此。设计学本身作为一种交叉学科，它推动了各类社会学科的创新发展。而作为一个新时代的设计学生，他们需要拓宽视野，探索涉猎学科的深度与广度，掌握新技术与新媒介的应用手段，才能够成为符合新时代背景的合格的设计师。

设计的目的是服务于人，也是实现人类追求美好生活的重要手段。设计的特征是集成创新，设计的目标是以需求为导向的转化应用。设计教育只有实施多向度的跨界、知识的交融、资源的整合、创新的集成、科学的评价，才能培养出能统筹多元知识、满足社会需求的合格的创新设计人才。

本套丛书是基于设计学学科的前期积累，综合了设计创新思维与方法、智能服装设计与教育、民族服饰与文化产业、民国图像与服饰历史、网络游戏与数字媒体、宜居城市广场群时空分布等研究成果，从多维度、多角度进行宏观与微观、传统与现代的多层面研究，努力丰富设计学学科的内容，拓宽学科视野。愿丛书的出版对设计学学科的发展起到积极的推动作用，与此同时，为高层次设计人才的培养以及设计教育范式转型与构建增添更多的理论支撑。

感谢本书所有作者同事们的大力支持。在编写过程中，疏漏之处在所难免，敬请各位同行及广大读者批评指正！

任文东

2020 年 8 月

PREFACE

The cultivation of interdisciplinary human talent in the field of fashion design is an issue of urgent importance both in the wider context of socioeconomic development and more specifically in terms of technological innovation and development. In terms of design education, developing interdisciplinary fashion design talent is very much part of a global education trend, which has arisen in response to the absolute demand for innovation and innovative thinking in all aspects of 21st century society. However, research into differing approaches to design in the fashion design sector is still at a developmental stage. Similarly, research that incorporates both design and new technology has been somewhat uncoordinated. The scholarly work in this field lacks an integrated philosophical approach which could serve to optimize and integrate relevant contributions from a diverse range of areas.

Bringing together a focused understanding of the fields of Knowledge Science (KS) and Fashion Design (FD), this thesis aims to establish a new strategy for interdisciplinary practical education called "Design Plus X" (DPX). This new strategy will be presented as a new and integrated approach to design thinking based on a crossover between multiple disciplines. The "X" of Design Plus X represents the knowledge that will be adopted and integrated from a number of non-design sectors and fields (e.g. sociology, management science, electronic engineering, information science, etc.). The purpose of this new practical research system DPX is to incorporate the knowledge and learning from various crossover departments and fields as they relate to design and the design process (the X component) into design. Ultimately, the hope is that new, comprehensive crossover multidisciplinary fields can be established such as Design Plus Humanities and Sociology, Design Plus Management and Design Plus Technology. This innovation will serve to drive forward the development of an integrated, innovative practical education.

The design process of DPX involves a series of four workshops and group presentations:
(1)Market: Identification of markets and main business drivers.
(2)Product: Generation of product feature concepts.
(3)Technology: Identification of technology solution options.
(4)Prototype: Refining or co-refining embodied solutions.

This process will form an iterative feedback loop that repeats until the desired outcome is achieved.

In terms of the concept of DPX a research approach that is humanities-oriented, will interact with fully conceived commercial strategies and detailed plans for plans for technological realization in order to present a distinctive and ground-breaking approach to integrated practical design education which includes a practical training component related to product design and services.

This thesis includes two case studies. The first is "Designing Comfortable Smart Clothing for Infant Health". The other is "A Fashionable Experience for Blind Children: Design Research into Intelligent Gloves with Chromatic Color Perception Function". The thesis has carried out extensive practical and experimental research in order to present the two case studies as showcases for the DPX thinking approach. By applying the DPX approach and philosophy, excellent design results have been achieved and a number of innovative products have been developed at the end of these projects.

The main qualities and the areas of significance of the approach outlined in this thesis can be briefly summarized as follows:

(1)The "Design Plus X" strategic thinking approach facilitates a new strategy for a practical and interdisciplinary fashion education. It offers an integrated new design thinking approach based on multidisciplinary crossover.

(2)During the design research development process of a project, "DPX", if applied consistently, can utilize learning from a whole range of multidisciplinary fields to produce a new integrated concept for design practice within practical education. The common issue of overemphasis on formal reorganization in design education can be resolved, and a core capacity for innovation centered on a new, revised approach to method and function can be cultivated.

(3)Based on the DPX approach, a fresh set of attitudes, aspirations, and capacities can be brought to the design education field. DPX provides a level of expertise that will set the standards for future work, thus contributing to the development of capable and creative designers. This innovative methodology encourages designers to approach problems and issues in their work in entirely new ways, utilizing entirely new perspectives. If the disciplinary boundaries between disciplines can be blurred or breached, there is a real opportunity to find practical and creative solutions to long-standing, seemingly intractable issues in the fashion design field. Overall, the whole philosophy of DPX aims at the leveraging of the knowledge from multiple domains in the wider objective of achieving positive societal outcomes.

(4)Based on the "DPX" thinking approach, an experimental Fashion Project was implemented. In the course of this action research, new methods were put forward in order to solve design problems. The results of this successful experiment are not only significant in themselves but they also provide great impetus for the further study of these proposed strategies for interdisciplinary practical fashion education.

(5)Utilizing a designer's perspective, we applied Knowledge Science to support research into a smart clothing design project. In turn, we promoted research and education in the field of KS based on knowledge from the Fashion Design and Smart Clothing Design fields. The research will make a significant contribution to the development of KS in moves towards a design for Human Life.

Ding Wei
July,2020

目 录
CONTENTS

Chapter 1
Introduction

Chapter 2
Literature Review

Chapter 3
Create New Thinking Approach for the Fashion Design Education Field

Chapter 4
The DPX Case Studies in Fashion Design Education: New Strategy for Interdisciplinary Practical Education with Specific Focus on Smart Clothing

Chapter 5
Discussions and Conclusions ·················· 131

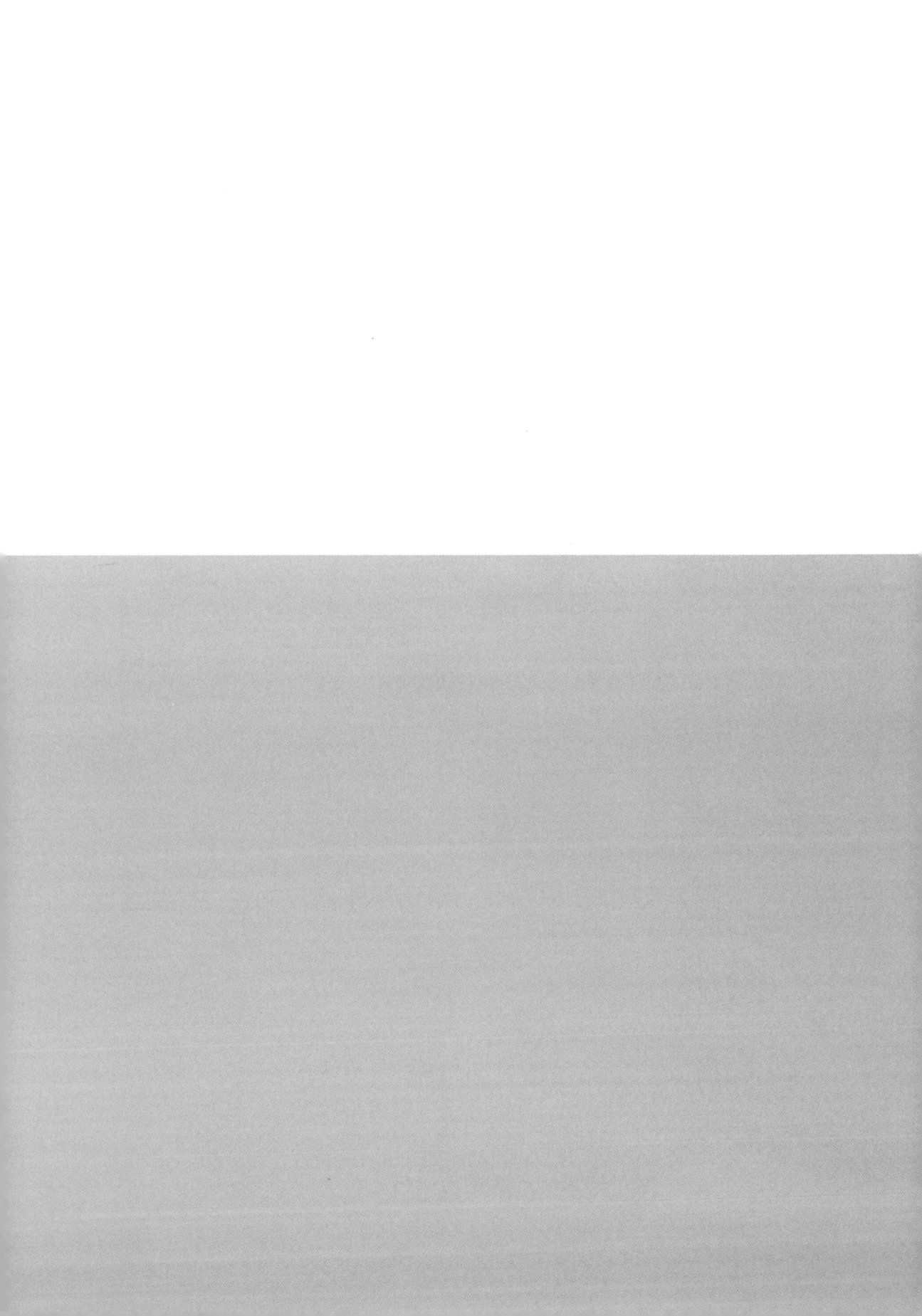

Chapter 1

Introduction

1.1

Research Background

1.1.1 The Demand Trends in the Fashion Market

With rapid technology development, the knowledge economy has made many influences on the fashion industry. Smart fashion has been paid lots of attention in the fashion industry. Many countries have made big efforts in the development and research of smart fashion. They are German, Japan, Finland, Swiss, the U.K, and so on. Many big companies have involved in developing and researching smart fashion. They are fashion companies, computer companies, or electronic application companies, such as IBM, Liwayisi, Phillips, Nick, and so on. Their smart fashion includes not only an appealing design but also wonderful functions. Their design could best satisfy the requirement of targeted customers. One of the American media in the field of technology predicted that future fashion would become a high-tech product with multiple functions. Many current design ideas would become reality. For example, it could record the heart rate and breath frequency. Moreover, it could play music, play video, adjust the temperature, and access the Internet simultaneously.

1.1.1.1 The Future of Fashion

What's next for fashion and tech? The fashion future is being described by trend analysts as 'post-human', which is inspired by artificial intelligence and illustrated by tech-driven design and a new generation of technical materials that are made from robust threads and metallic filaments (Heather Picquot 2017). Properties of post-human apparel are ultra-light, highly breathable, protective, and thermo regulating (capable of regulating your body's temperature), while futuristic fashion is defined by otherworldly aspects like embossed textures, bespoke fabrics, liquid or metallic finishes, and glowing LED lights inside of mesh materials. Futurism is further supported by the rise of 3D-printed clothing, footwear, and accessories, led by brands like The Unseen, Nike, and Intel.

The Future of Fashion and Tech: 4 Fascinating Predictions (Fashion Snoops 2017).

(1)Wearables are becoming much more comfortable and fashion focused. A great example is Samsung's Body Compass that was unveiled at CES 2016 in January. It's a tank top with a tiny visible metal disk, a bottle cap - sized sensor. The compass is complete with a battery and sensors beneath the fabric that can measure heart rate, stance and body fat levels.

(2)Using technology to get deeper in touch with our human senses is becoming an apparent man date of man new tech/textile companies. Doppler Lab's Here earbuds took home the Best of Show Award at SXSW Interactive for their optimized listening technology that changes the way we hear the world. In the scent area, carbon alloy textiles absorb odors and block scents, while cosmetic textiles have the ability to release scents among other micro-encapsulation technologies like cooling and moisturizing. Even areas of pain can be eliminated with Quell's Velcro band that uses neuro technology to block areas of chronic pain by triggering sensory nerves.

(3)The growing industry of 3D printers and open-source technology gives every one of us the ability to be fashion innovators. We love the work that Created a technology and textile studio, has done including the heated jacket, a programmable LED matrix dress for Zac Posen, and more.

(4)Sustainability and social responsibility are quickly becoming requirements for fashion brands. Advancements in technology play pivotal roles here, from waterless dyeing techniques to Stella McCartney's use of cutting-edge sustainable materials into her collections by pushing the boundaries of what a sustainable product can look like, including from her synthetics that replace leather and biodegradable soles made from a bioplastic called APINAT.

1.1.1.2 Existing Wearables and Smart Clothing Market Landscape and Market Size

(1)Existing Wearables and Smart Clothing Market Landscape.

With the rapid technology development, knowledge economy has made many influences on fashion industry. Innovating design by using high knowledge has become the best solution and new trend for this industry. High technology and information technology are the basis of knowledge economy. The main task of fashion industry is to improve the functions of traditional garment by applying high technology and information technology from now.

Recently, wearables and smart clothing have been paid lots of attentions in fashion industry. Many countries have made big efforts in the R&D of smart fashion. They are USA, German, Japan, Finland, Swiss, the U.K and so on. When it comes to new period of wearable computers, both big companies and startups are experimenting new features and developments. Their smart fashion includes not only appealing design but also wonderful functions. Their design could best satisfy the requirement of targeted customers. The main goal of wearable technology is to design stylish and invisible wearable devices which leading to much competition in the market. Considering that wearable technology is a relatively new industry, the landscape is populated with vendors across many markets. Companies in the wearables market have gravitated to four primary marketplaces（Figure 1-1）：

①Infotainment real-time data transmission for entertainment.

②Fitness and wellness monitoring of activity and emotions.

③Military and Industrial real-time data transmission in military or industrial environments.

④Healthcare and Medical monitoring of vital signs and sense augmentation.

The Table 1-1 shows the products that are offered in the four leading market categories.

Wearable devices do have wearable sensors embedded in it which captures body activities like blood pressure, heartbeat ,etc. attracting the health industry (Pranay Sni, 2014). The usage of wearable sensors is increasing due to growth in health industry and people awareness towards health and fitness. The wearable devices used in the defense area are smart glasses and smart clothing. There are many big names associated with new

Figure 1-1 Existing wearables and smart clothing landscape

Source: IHS Electronics & Media. An IHS whitepaper: wearable technology-market assessment[R].2013,09.

Table 1-1 Existing wearables and smart clothing landscape

Category	Product offerings	
Infotainment	smart glasses virtual reality goggles heads−up displays	smart watches bluetooth headsets
Fitness and Wellness	smart clothing activity monitors fitness and heart rate monitors pedometers	sleep sensors smart glasses smartwatches emotional monitors
Military and Industrial	smart clothing hand−worn terminals	heads−up displays smart glasses
Healthcare and Medical	smart clothing biometric monitors chemical monitors drug delivery products	smart glasses hearing aids smart watches defibrillators

Source: IHS Electronics & Media. An IHS whitepaper: Wearable Technology Database | Vandrico Inc [R]. 2016,04.

developments occurring in wearable devices such as Apple, Accenture, Nike, Samsung and maybe more (Figure 1-2).

(2) Existing Wearables and Smart Clothing Mark Size.

The market of wearable technology is expanding with a high rate gradually. There is the expectation of crossing the business of US$8 billion by the end of 2018 with CAGR increasing by 17.7% from the year 2013 to 2018 widely in the world (Figure 1-3). Technologies which are on arrival are Apple testing iWatch, iRing, S6 Golf Watch, Wrist Gear, Sony Smart Band, 3 High-Tech Eye Glasses, Google Glass, Bluetooth Ring, iPhone-Connected Jewelery with wireless security alerts, Smart Contact Lenses (Smita Jhajharia, 2014) (Figure 1-4).

The 2015 global smartphone market is an impressive $399 billion, but pales in comparison to the clothing market with $1.2 trillion in garment sales. For 2019, this gap is predicted to widen to $520 billion in smartphone sales and a whopping $2.2 trillion in garment sales. Given the pervasiveness and continual growth of the clothing market, you would assume that merger of wearable technology with clothing would be an obvious area for market expansion. However, growth in

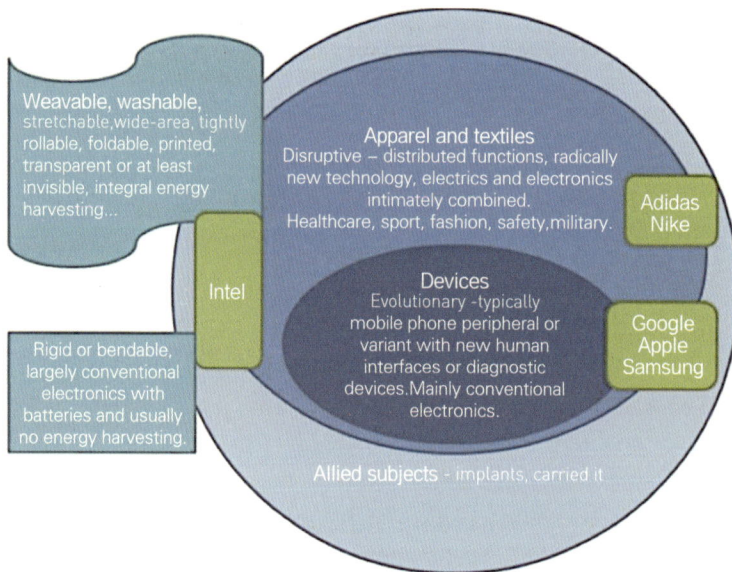

Figure 1-2 The two main types of wearable technology and their typical characteristics
Source:P Harrop, J Hayward, R Das, G Holland. Wearable Technology, 2015.

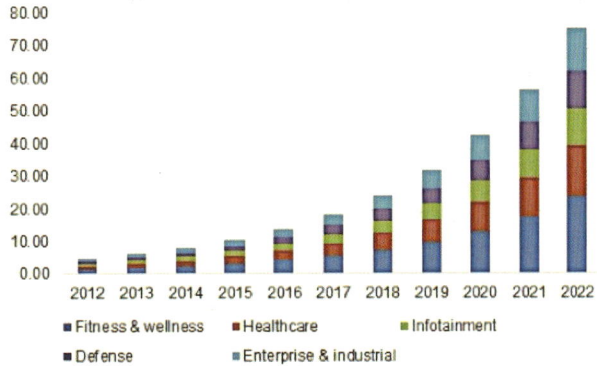

Figure 1-3 Wearable technology market analysis-size, share, growth, trends and forecasts to 2022

Source: M Mahimkar. Wearable Technology Market Analysis-Size, Share, Growth, Trends and Forecasts To 2022. 2016.

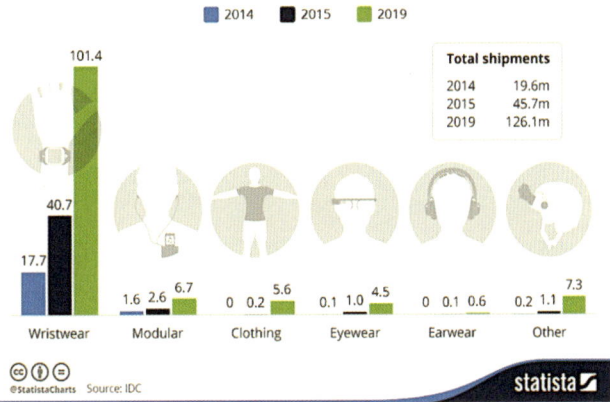

Figure 1-4 Prediction of wearables

Source: M Nguyen. The most successful wearables. 2016.

this area is predicted to be slow, with smart clothing account for less than 1% of the market (Figure 1-5, Figure 1-6).

Nowadays, science and technology have entered a new era. Intelligent digitization has become the theme of the fashion design. Technology drives

Cumulative Wearable Device Shipments by Device Category, World Markets: 2013-2020

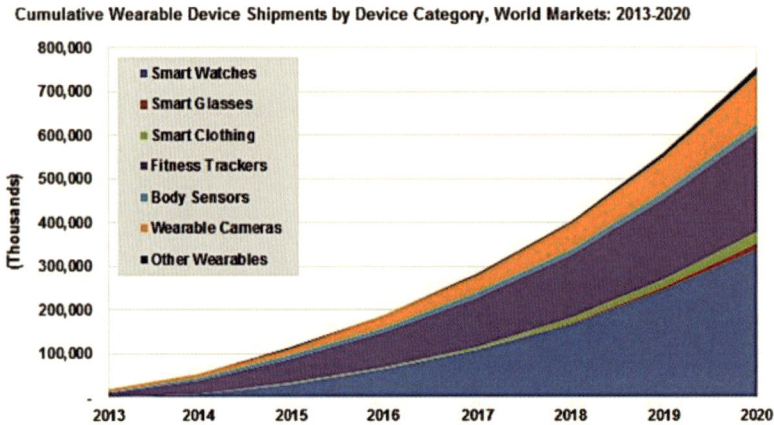

Figure 1-5 Smart clothing market size

Source:Cumulative Wearable Device Shipments to Surpass 750. 2015, 18 April 2016.

Level of Familiarity With Wearable Devices

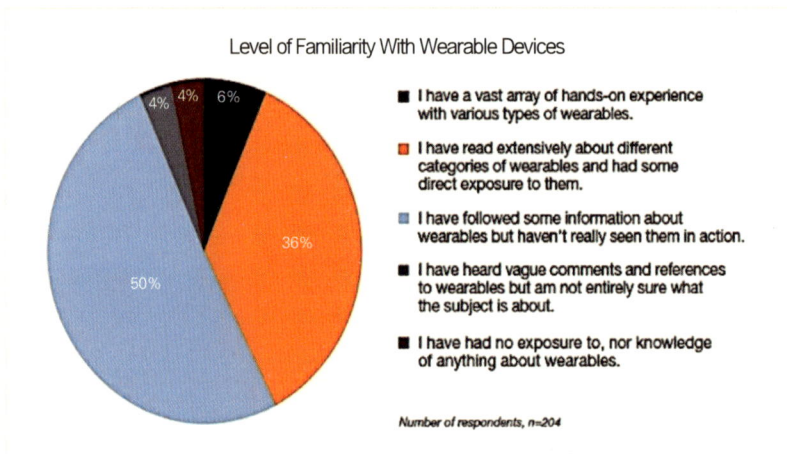

Figure 1-6 Level of familiarity with wearable devices

Source: T Maddox. Wearables: Fit For Business. 2016.

the rapid development of clothing art. The garment has got a qualitative leap. The purpose of the fashion design is not only to pursue fashion and aesthetic. It is a multi-disciplinary comprehensive design art. Grasping the frontiers of modern garment design art, we need to explore a new perspective of modern art of clothing, starting from the multi-angle and multi-dimensional unceasingly, communicating with multi-culture and setting up a new conception of taking health as the center all that are to lead the fashion design to a broader field with the backing of a strong technical force. Smart clothing design area has brought new hopes and opportunities in various field leading to development all over the world.

Big data analysis is necessary to realize the full potential of smart clothing. There are many smart clothing products and use cases where providing personal biometric or environmental data to end users can improve the health, safety, and happiness of consumers. However, these use cases have limited market value and will probably have limited growth in the next 5 to 10 years (Alex Hanuska & Bharath Chandramohan, et al. 2017).

Companies who develop products that use big data to identify trends and can use the predictive analytics to improve the performance, health, and safety of larger populations will achieve market success. With the larger wearable industry driving technical innovation in sensor size and capability as well as a new market of IoT platforms, smart clothing is positioned to have the tools for market success. By focusing on commercial application, cultural challenges of style and expense become less of a barrier. Although consumer wearable and smart clothing technologies have larger public attention, business and enterprise application will give the market financial value that drive growth in years to come.

1.1.2 The Current Situation of Fashion Design Education

The necessity for highly specialized human talent has newly emerged due to technological, economical, and societal developments, which is the driving force propelling changes in higher education of fashion, while at the same time, indicating the direction of reform and development within fashion design education. Under new circumstances consisting of advanced technological integration, the daily increasing complexity of social systems, and ferocity of international competition, higher education in fashion is pressed by the need to transcend the boundaries of single-major education, do away with disordered mixing and factionalism of academic departments, adhere to the principle of growth of innovative human talent, and foster interdisciplinary human talent for a mass scales who are adapted to contemporary needs .

The cultivation of interdisciplinary human talent in the field of fashion design is an urgent issue for technological innovation development and socioeconomic development, while also conforming to the principle of growth of innovative human talent, and becoming a global trend towards reform in postgraduate education. Norbert Wiener, the proponent of cybernetics, once correctly pointed out the following: 'There is a great fertilizing and revivifying value in the contact of two scientists with each other, but this can only come when at least one of the human beings representing the science has penetrated far enough across the frontier to be able to absorb the ideas of his neighbor into an effective plan of thinking'. In recent times, scientific crossover and integration have become a source of technological innovation. Technological development requires multiple specialists of different fields to work collaboratively, and furthermore requires the existence of large numbers of highly specialized human talent with an interdisciplinary perspective and thinking, who have acquired theories and methods from multiple fields, and possess the ability to study and refer to the results of other specialist fields. As fashion design education is an important route for highly specialized human talent, the pursuit and establishment of a new model for cultivating interdisciplinary human talent is an inevitable requirement for technological innovation and development. Such human talent can, using knowledge and methods from multiple fields, deal appropriately with complex issues of various types, and effectively propel innovation in knowledge, technology, and systems in order to freely respond to kaleidoscopically changing markets.

Therefore, whether it be solutions to complex issues directly faced within socioeconomic development, or adjustment to integrated professions or jobs, in any event, the cultivation of interdisciplinary human talent in higher education, in particular, postgraduate education is necessary. A look at the situation in the USA, the UK, Germany, and Japan shows us that the cultivation of interdisciplinary human talent in the design field has become an important trend in reform in global postgraduate education reform.

1.1.3 Summary

With rapid technology development, the knowledge economy has made many influences on the fashion industry. It would shape the character of the development of fashion design in the 21st century. Innovating design by using high knowledge has become the best solution and new trend for this industry. Fashion could be looked like one of the vehicles of modern technology. High technology and information technology are the basis of the knowledge economy. The main task of the fashion industry is to improve the functions of traditional garments by applying high technology and information technology from now. The future of fashion is in tech. The fashion industry needs real leaders, who are ready to tap into the powerful opportunities of evolving technologies.

1.2

Research Questions

The research hypothesis posed is based on a set of research questions. Donald Norman describes the key issues in a recent article on the changes required for design education today. In the early days of industrial design, the work was primarily focused upon physical products. Today, however, designers work on organizational structure and social problems, on interaction, service, and experience design. Many problems involve complex social and political issues. As a result, designers have become applied behavioral scientists.

They are woefully undereducated for the task. Designers often fail to understand the complexity of the issues and the depth of knowledge already known. They claim that fresh eyes can produce novel solutions, but then they wonder why these solutions are seldom implemented, or if implemented, why they fail. Fresh eyes can indeed produce insightful results, but the eyes must also be educated and knowledgeable. Designers often lack the requisite understanding. Design schools do not train students about these complex issues, about the interlocking complexities of human and social behavior, about the behavioral sciences, technology, and business. There is little or no training in science, the scientific method, and experimental design(Donald Norman,2012).

The emergence of these new problems brings new challenges in the field of Fashion Design education. As the conventional Fashion Design education and clothing design process is no longer suitable for the social needs, it's essential to conduct theoretical innovation and gradually sum up a set of suitable for Fashion Design education and design practice.

The research questions culminate into a hypothesis. How to apply the knowledge of knowledge science to support fashion design education and research work? Which is as follows: Could we combine Knowledge Science and Fashion Design these two knowledge field to creative an integrated thinking approach to optimize the input from the different areas. And the new thinking approach is intended as an interdisciplinary education strategy to guide the education research and design research.

1.3

Research Aim

Combining knowledge Science (KS) and Fashion Design (FD), The research aims to establish a new strategy for interdisciplinary practical education called "Design Plus X", as an integrated new design thinking approach based on the multidiscipline crossover. "X" corresponds to knowledge from various crossover departments and fields (e.g. sociology, management science, electronic engineering, information science, etc.). The purpose of the new practical system research called Art Design DPX is, with the whole process of design development as a carrier, to incorporate the knowledge from various crossover departments and fields represented by X into the design, utilize the broadening and deepening of knowledge related to design development, to comprehensively crossover multidisciplinary fields such as Design Plus Humanities and Sociology, Design Plus Management and Design Plus Technology, and to develop an integrated, innovative practical education. This aim is then further expanded into:

(1)To understand knowledge Science education and main knowledge theories of KS.

(2)To understand Fashion Design education and main design methodology of FD.

(3)To understand main existing conceptual models in Fashion Design, Knowledge Science and Smart Clothing Design (SCD).

(4)To study how to incorporate integration of two knowledge areas KS and FD, then create an integrated thinking approach (Design Plus X) to optimize the input from the different areas.

(5)To study from action research: Case study (1) Designing Comfortable Smart Clothing: for Infants' Health Monitoring. Case study (2) Fashionable Experience for Blind Children—Design: Research of Intelligent Glove Featuring Perception of Chromatic Color. To through study on the theory and practice of Smart Clothing to identify appropriate new thinking approach, encourage designer from different way to thinking problem, mix knowledge break discipline bound to solve issues of fashion design field.

(6)To promote research and design education in Fashion design and Smart clothing design based on the field of Knowledge Science.

1.4
Scope of Research

The research focuses on four areas: Knowledge Science Education, Fashion Design Education, Smart Clothing Design Development Process, Smart Technology (Figure 1-7).

1.4.1 Knowledge Science

Knowledge science is a problem-oriented interdisciplinary field that takes as its subject the modeling of the knowledge creation process and its application, and carries out research in such disciplines as knowledge management, management of technology, support for the discovery, synthesis and creation of knowledge, and innovation theory with the aim of constructing a better knowledge-based society (Nakamori, 2011).

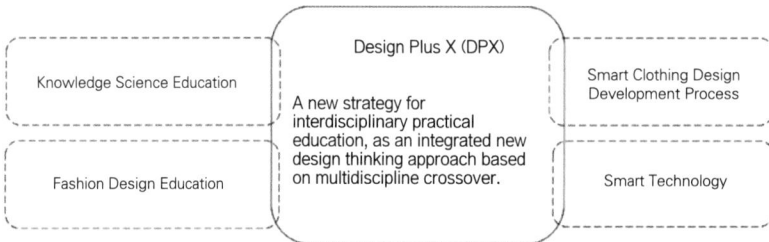

Figure 1-7 Diagram showing focus of the research

Knowledge science has different way with other academic discipline in strengths and features. The essence of knowledge science is how to recombine resource and knowledge-structure. It is also the essence of innovation. Knowledge science with a bird's-eye view on the advent of knowledge-based society, and the spread of knowledge management. It can guide researcher to use the creative thinking methodology to make something new and design something to face the future. We apply the knowledge of knowledge science to support smart clothing design research work from designer perspective, promote research and design education in Fashion design and Smart clothing design based on the field of Knowledge Science.

1.4.2 Fashion Design

Fashion design is the art of application of design and aesthetics or natural beauty to clothing and accessories. Fashion design is influenced by cultural and social attitudes, and has varied over time and place. Fashion designers work in a number of ways in designing clothing and accessories such as bracelets and necklaces. Nowadays, entering a new era of science and technology. Intelligent digitization have become the theme of the fashion design. Technology drives the rapid development of clothing art. The garment has been a qualitative leap. The purpose of the fashion design is not only the pursuit of fashion and aesthetic. It is a multi-disciplinary comprehensive design art. Grasping the frontiers of modern garment design art, we need exploring a new perspective of modern art of clothing, starting from the multi-angle and multi-dimensional unceasingly, communicating with multi-culture and setting up a new conception of taking health as the center all that are to lead the fashion design to a broader field with the backing of a strong technical force.

1.4.3 Smart Clothing Design

Smart clothing is a "smart system" capable of sensing and communicating with environmental and the wearer's conditions and stimuli. Stimuli and responses can be in electrical, thermal, mechanical, chemical, magnetic, or other forms (Tao, 2001). As the topic of smart clothing belongs to the interdisciplinary studies covering computer technology and clothing design, such an area of research aims to extend the functionality of traditional fabrics and apparels by means of computer and materials science and technology. It can make this new style of clothing better suited to people in the specific work environments, such as keeping the human warm and avoiding injuries, and even monitoring people's health and physical conditions in variation. On the one hand, the computer and material scientists highlight the effective integration of sensors and other electronic components into the fabrics, so that the fabrics can effectively sense the variations of temperature, humidity, current and other parameters in the surrounding environment. On the other hand, the apparel designers shall be engaged in the design that suits a particular environment and customers, depending on the specific fabrics and the roles of apparels. In order to achieve excellent design, the apparel designers need to take into account the features of fabrics, the characteristics of electronic components, the aesthetic appearance, comfort and safety of fabrics, the roles and aesthetics of apparels and other factors. This field as a new continent in the apparel industry has huge demands and potential market, however, it also shows a number of new issues with challenges.

1.4.4 Smart Technology

Electronic technique, Sensor technology, Communication technology, Wireless sensor networks (WSN), GPRS, Bluetooth, Micro-processing, etc.

1.5

Research Methods

(1)Theory on knowledge creation.

I-system: A knowledge construction model.

(2)Case study by Design Plus X (DPX) thinking approach.

The method of clothing design, the principle of comfort quality in fashion design, ten principles for good design, smart technology.

(3)Reference.

Include textbook, journal, encyclopedias and collections of online reference books.

(4)Opinion.

Includes views, judgments or appraisals of individuals or design team, the techniques include questionnaires, brainstorming.

(5)Laboratory.

Experimental design. It has a structure and method of collecting and analyzing: Machine learning algorithms and Python programming language.

1.6

Research Contributions

This research aims to provide five key contributions.

(1)New strategy.

The "Design Plus X" strategical thinking approach facilities new strategy for interdisciplinary practical fashion education, it offered an integrated new design thinking approach based on multidiscipline crossover.

(2)New integrated concept.

"DPX" with the whole process of design research development as a carrier, multidisciplinary fields will crossover comprehensively, and by enforcing the consistency of Design Plus X, a new integrated concept for design practice within practical education, overemphasis on formal reorganization in design education can be resolved, and a core capacity for innovation that centers on reform of method and function will be cultivated.

(3)New standards.

Based on DPX thinking approach, it brings a fresh set of attitudes, aspirations, and capacities. It provides the expertise, sets new standards that others will rise to, and contributes to the development of capable and creative designers. It is encourage designer from different way to thinking problem, mix knowledge break discipline bound to solve issues of fashion design field. The integration of multiple domains is encouraged in order to respond to the need for societal wellbeing.

(4)New methods.

Based on "DPX" thinking approach, we design the experiment of Fashion Project. In the action research, we provide the new methods to solve the problem of design. This not only constitutes a significant achievement but also encourages further study of the proposed strategy for interdisciplinary practical fashion education.

(5)New perspective.

We apply the KS to support smart clothing design research work from designer perspective, promote research and education in the field of Knowledge Science based on Fashion Design and Smart Clothing Design. The research works will contribution to develop KS into design for Human Life.

1.7

Structure of the Thesis

This thesis consists of five chapters as shown below. The contents of each part are:

(1)Chapter 1: introduction.

This chapter introduces the background to the research during the construction of the research proposal including the hypothesis, research questions, rationale of the research, with aims of the project, research methods and research key contributions.

(2)Chapter 2: literature review.

This chapter introduces three key subject areas, namely Knowledge Science education, Fashion Design education, Smart Clothing development (include smart clothing design process and smart technology). Summaries the key problems and hypotheses, and explains the purpose of this research including goal and objectives.

(3)Chapter 3: create new thinking approach for the fashion design education field.

This chapter focuses on the theoretical basis of the research, which leads to the contribution of new knowledge. Combining the knowledge of Knowledge Science and Fashion Design, Aim to establish a new strategy for interdisciplinary practical education called "Design Plus X", as an integrate new design thinking approach based on multidiscipline crossover. "X" corresponds to knowledge from various crossover departments and fields. The purpose of the new practical system research called Art Design DPX is, with the whole process of design development as a carrier, to incorporate the knowledge from various crossover departments and fields represented by X into design, utilize the broadening and deepening of knowledge related to design development, to comprehensively crossover multidisciplinary fields such as Design Plus Humanities and Sociology, Design Plus Management and Design Plus Technology, and to develop an integrated, innovative practical education. The design process of DPX involves a series of four workshops and group presentation:market, product, technology, prototype.This is an iterative feedback loop that takes place until the desired outcome is achieved.

(4)Chapter 4:The DPX case studies in fashion design education: new strategy for interdisciplinary practical education with specific focus on smart clothing.

This chapter introduces two case studies, one project is" designing comfortable smart clothing: for infants' health", the other one is "fashionable experience for blind children——design research of intelligent glove featuring perception of chromatic color". This chapter documents the experimental stage of the practical research and explains the two case studies employed the DPX thinking approach to create the new production and got the good design outcomes.

(5)Chapter 5: discussions and conclusions.

This last chapter provides a summary and discussion about the whole research including methods, findings and outcomes, and the recommendation on how to understand the DPX thinking approach and this research.

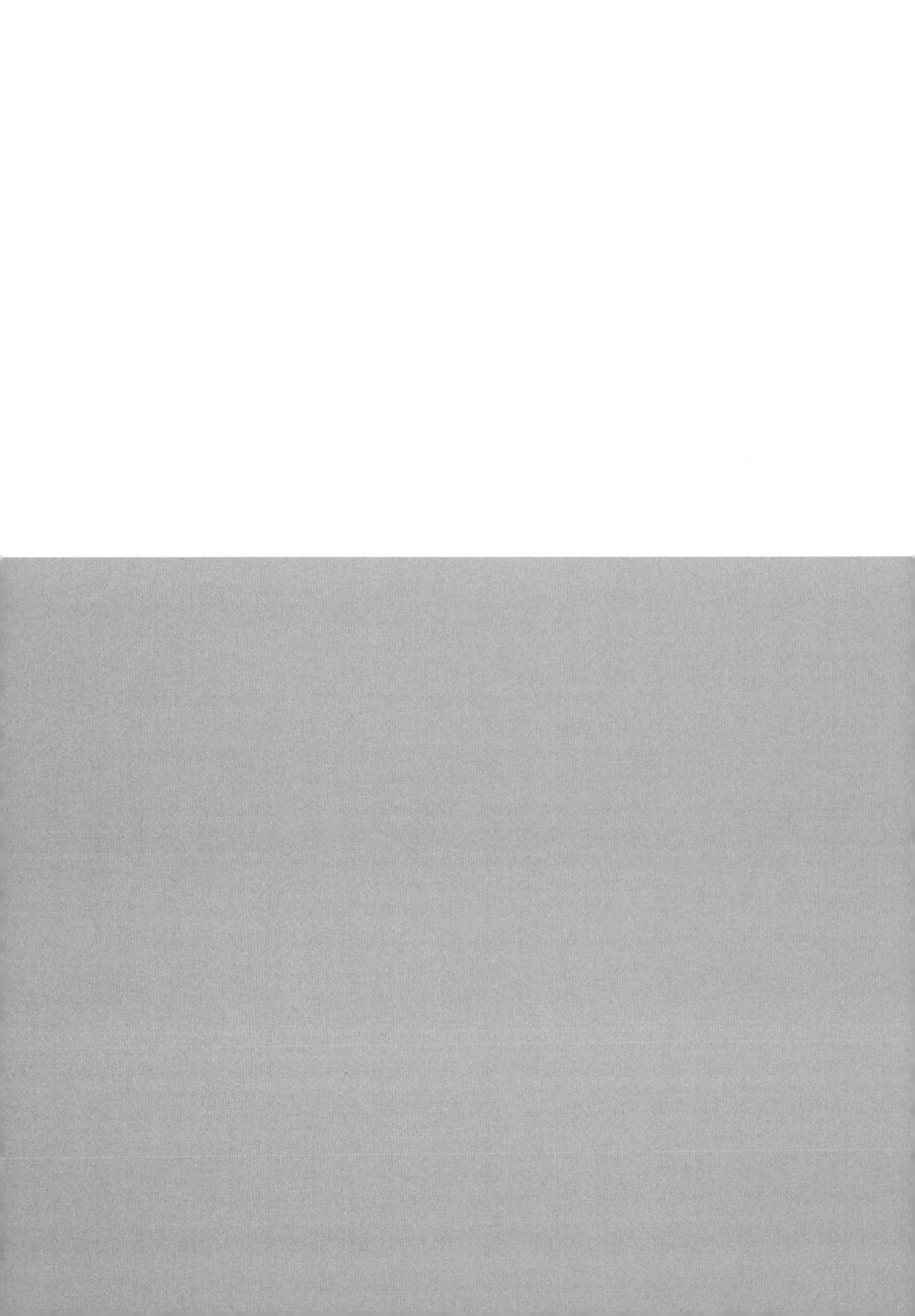

Chapter 2

Literature Review

2.1

Overview of Knowledge Science Education

The School of Knowledge Science at Japan Advanced Institute of Science and Technology (JAIST) started education in April 1998, which is the first school established in the world to make knowledge a target of science. The first dean of the School was Professor Ikujiro Nonaka who is famous worldwide for his organizational knowledge creation model called the SECI spiral (Nonaka & Takeuchi, 1995), which is in fact the key factor in establishing the School.

2.1.1 Definition of Knowledge Science

Knowledge science is a problem-oriented interdisciplinary field that takes as its subject the modeling of the knowledge creation process and its application, and carries out research in such disciplines as knowledge management, management of technology, support for the discovery, synthesis and creation of knowledge, and innovation theory with the aim of constructing a better knowledge-based society (Nakamori, 2011). Knowledge science is a scientific discipline that draws on Philosophy, Management, Information and Systems sciences' methodologies along with emerging technologies in Neuroscience and Complexity sciences to, first of all, study what "Knowledge" is, and to establish and develop principles, theories and methodologies of how knowledge can be created, managed, synthesized and used (Figure 2-1).

At the present stage, knowledge science is more a problem-oriented interdisciplinary academic field than a single discipline. Its mission is to organize and process both objective and subjective information and to create new value and knowledge. Knowledge science mainly deals with the research area involving social innovation such as regeneration of organizations, systems and the mind. However, society's progress is underpinned by technology, and the joint progress of society (needs) and technology (seeds) is essential. Therefore, knowledge science also has the mission to act as a coordinator (intermediary) in extensive technological and social innovations.

In order to fulfill these missions, the School of Knowledge Science focuses its research and education on observing and modeling the actual process of carrying out the mission, as described in the organizational knowledge creation theory by Nonaka and Takeuchi (1995), or the creative space theory by Wierzbicki and Nakamori (2006), as well as developing methods to carry out the mission. The methods are mainly being developed through the existing three areas in the School. They are:

(1)The application of business science/organizational theories (practical use of tacit knowledge, management of technology, innovation theory).

(2)The application of information technology/artistic methods (knowledge discovery methods, ways to support creation, knowledge engineering, cognitive science).

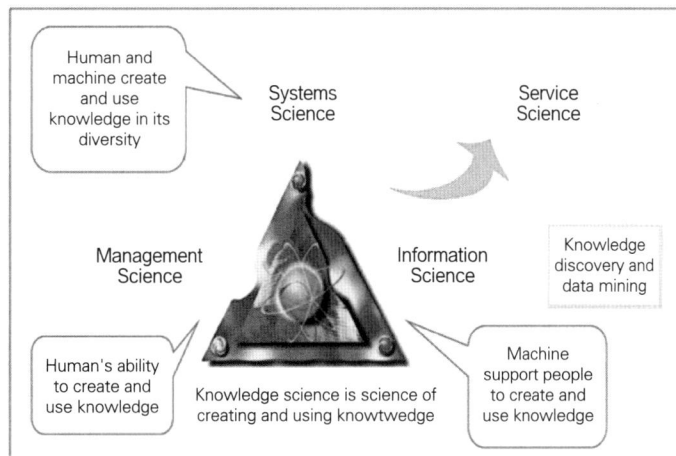

Figure 2-1 Knowledge science

(3)The application of (mathematical) systems theory (systems thinking, the emergence principle, socio-technical systems).

2.1.2 Approaches to Knowledge Science

We could count several research fields related to knowledge science:

(1)Knowledge engineering symbolizes (approximates) experts' knowledge to develop artificial intelligence.

(2)Knowledge discovery mines a large scale of data set to extract partial rules, and adds their meanings, using domain knowledge.

(3)Knowledge construction simulates complex phenomena based on some hypothesis, and adds the meanings of emerged properties, using domain knowledge.

(4)Knowledge management tries to convert distributed (or tacit) knowledge into shared (or explicit) knowledge, and uses it effectively.

However, to solve complex real-life problems we need knowledge synthesis, collecting and interpreting different types of knowledge from cognitive-mental front, scientific-factual front, and social-relational front. We have been trying to establish a theory related to knowledge synthesis because we believe that the most important task is knowledge synthesis in knowledge science as well as systems science. We have just published a first version of the theory in Systems Research and Behavioral Science (Nakamori, et al., 2011).

Nonaka et al. (2000) called the dynamic context which is shared and redefined in the knowledge creation process "Ba", which does not only refer to a physical space, but also includes virtual spaces based on the Internet, for instance, more mental spaces which involve sharing experiences and ideas. They stated that knowledge is not something that can exist independently; it can only exist in a form embedded in 'Ba', which acts as a context that is constantly shared by people.

From the hypothesis that knowledge science will be established at the "Ba" where three disciplines are integrated, we should expand our research into social and technological innovation to foster revitalization projects and collaborative projects with enterprises.

2.1.3 Main Knowledge Theories

2.1.3.1 Data, Information and Knowledge

Data science can be seen as the study of the generalizable extraction of knowledge from data. Data Scientist: The Sexiest Job of the 21st Century (Harvard Business Review, 2012). "in God we trust, all others bring data" (W.E.Deming). Data science is used interchangeably with data analytics knowledge discovery and data mining emerged as a rapidly growing interdisciplinary field that merges together databases, statistics, machine learning and related areas in order to discover and extract valuable knowledge in large volumes of data.

The question of defining knowledge has occupied the minds of philosophers since the classical Greek era and has led to many epistemological debates. Epistemology is the study of the nature and grounds of knowledge. Epistemologists reason that knowledge is justified true belief. They contemplate the eternal challenge of separating true from false. As Nonaka and Takeuchi point out: "we consider knowledge as a dynamic human process of justifying personal belief toward the truth." A line of thinking in this branch of philosophy recognized knowledge as awareness of absolute and permanent facts. Kantian synthesis, a foundation of modern rationalism and empiricism, later resulted in the notion that knowledge expresses an organization of perceptual data on the basis of diverse categories; these categories included space, time, objects and causality. The subjectivity of basic concepts about space and time, discovered by mathematics and physics in 19th and 20th century, has changed essentially epistemology. At the turn of these centuries, an American movement in philosophy called pragmatism, founded by C.S.Peirce and William James, extended the definition of knowledge; later, it was further modified as a result of the influence of artificial intelligence and quantum mechanics. It stated that knowledge consists of models, which reflect the surrounding environment, resulting in a targeted, simplified problem-solving and cognitive conclusions. The evolution of epistemology resembles the modern recognition of the knowledge hierarchy: data to information to knowledge, and further to wisdom as represented in Figure 2-2. That is the Data Information Knowledge and Wisdom (DIKW) hierarchy.

Data are raw facts and numbers, which can be informative but by themselves provide little value for decision making, planning, or any other actions. Data gain meaning once they are put into context and once in the relations between data and context are understood. "When endowed with relevance and purpose," as Drucker stated, "data become

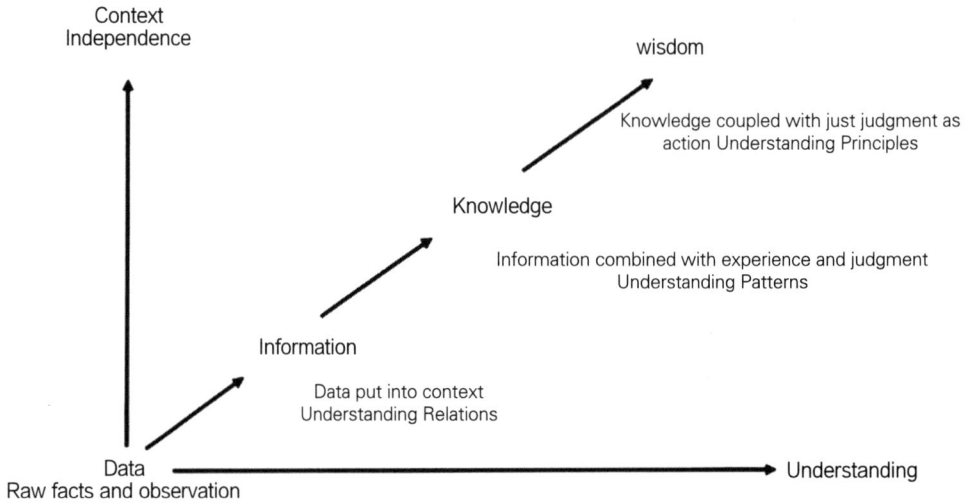

Figure 2-2 The Data Information Knowledge and Wisdom (DIKW) hierarchy

information". Davenport and Prusak listed the following values added by the transformation of data into information: contextualization, categorization, calculation, correction and commencement. Knowledge combines information with individual, group and organizational experience and judgment, and it involves making a leap from understanding relations to patterns that can guide actions. Davenport and Prusa point to the following processes involved in the transformation of information into knowledge: comparison, consequence, connection and conversation. Zeleny also proposes additions to the DIKW hierarchy. According to him "enlightenment" should be on the top of the familiar DIKW framework. Enlightenment, according to Zeleny (personal communication, 29 October,2004) "is not only answering or understanding why (wisdom), but attaining the sense of truth, the sense of right and wrong, and having it socially accepted, respected and sanctioned."

However, the assumption of a hierarchy from data to information to knowledge with each varying along some dimension, such as context, usefulness, or interpretability, does not survive a critical evaluation, The hierarchy is problematic because it is difficult to distinguish between data and information and between information and knowledge. Zack argues that data can be considered as facts or observations whereas information is data in a context; knowledge is information that is accumulated and organized in a meaningful way. Rather, knowledge is information possessed in the mind of individuals: it is personalized information (which may or may not be new, unique, useful or accurate) related to facts, procedures, concepts, interpretations, ideas, observations and judgments. Tuomi makes the iconoclastic argument that the often-assumed hierarchy from data to knowledge is actually inverse: knowledge must exist before information can be formulated and before data can be measured to form information. As such, raw data do not exist, even the most elementary piece of "data" has already been influenced by the thought or knowledge processes that led to its identification and collection. Tuomi argues that knowledge exists which, when articulated, verbalized and structured, becomes information which, when assigned a fixed representation and standard interpretation, becomes data. Critical to this argument is the fact that knowledge does not exist outside of an agent (a knower): it is indelibly shaped by ones needs as well as one's initial stock of knowledge; knowledge is thus the result of cognitive processing triggered by the inflow of new stimuli. Consistent with this new view, Alavi and Leidner state that information is converted to knowledge once it is processed in the mind of individuals and knowledge becomes information once it is articulated and presented in the form of text, graphics, words or other symbolic forms. Braganza also proposed a Knowledge-Information-Data (KID) (Figure 2-3) model based on a case study, which suggested knowledge leads to information which determines data. It reserved the commonly accepted hierarchy which assumes that knowledge is a product of data and information.

2.1.3.2 Knowledge Creation Processes

Knowledge Creation is a dynamic activity that can enhance organization success and economic well-being. It is an important driver of innovation, which is not only for organizational competitiveness and survival, but also can have far-reaching societal, notional and global consequences.

Since the last decade of the 20th century, in quite new approaches to knowledge creation. The first of such approaches is the Shinayakana Systems Approach. Influenced by the soft and critical systems tradition, it specifies a set of principles for knowledge and technology creation. To these principles belong: using intuition, keeping open mind, trying diverse approaches and perspectives including all advancements of both hard and soft systems science, being adaptive and ready to learn from mistakes, being

Knowledge can be considered data at a
high level of abstraction and generalization.

Obtaining by
– Perceiving
– Discovering
– Learning

Obtaining by
– Processing

Obtaining by
– Observing
– Measuring
– Collecting

Knowledge

Information

Data

Integrated information, including
facts and their relations
(justified true belief).
Is this road appropriate for such
amount of cars?

Data equipped with
meaning.
Average of number of cars
each hour, each day, each
week, each year on the road.

Un–interpreted signal
Number of cars
counted on a road by
hours, by days of the
week, by months.

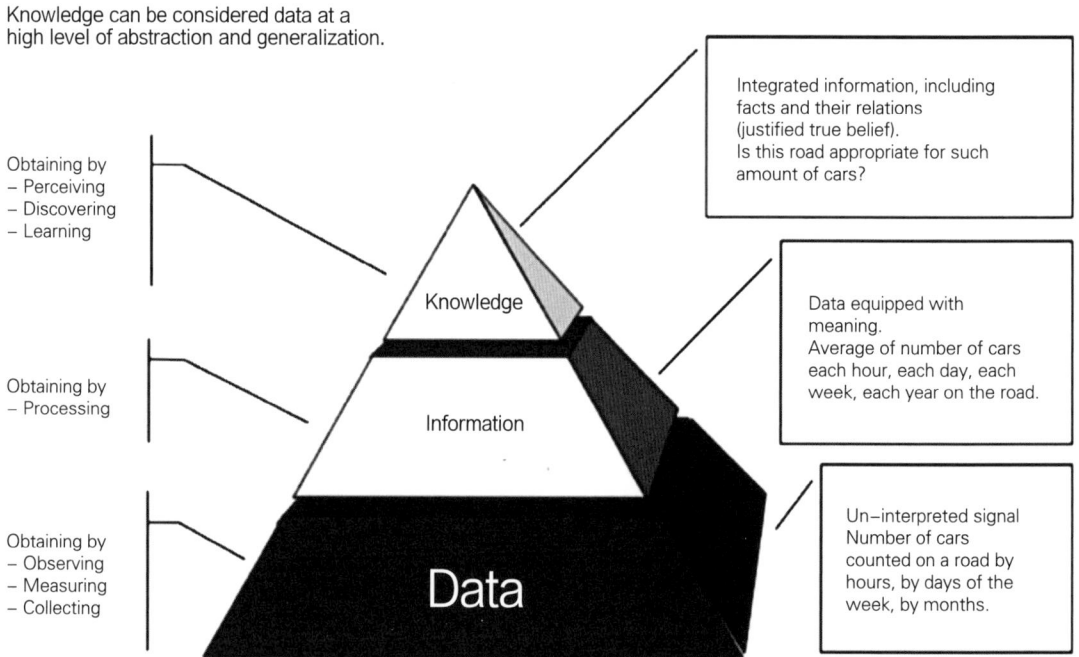

Figure 2-3 Knowledge can be considered data at a high level of abstraction and generalization (KID)

elastic like a willow but hard as a sword. Another systemic and process-like approach to knowledge creation, called I-system, was developed based on the Shinayakana Systems Approach. The five ontological elements of this system are Intervention (problem and requirement perspective), Intelligence (public knowledge and scientific dimension), Involvement (social motivation), Imagination (creative dimension), and Integration (synthesized knowledge), as Figure 2-4 and Figure 2-5 shown. All transitions between diverse dimensions of creative space are free according to individual needs. The I-system also appears sophisticated and inherently reflection in sociology view. Nakamori and Zhu further explored I-System as a re-structuration model for knowledge construction from sociological point. Viewed through I-system, knowledge, as emergent and continuous social practice (Intervention) and accomplishment (Integration), is constructed by actors, who are constrained and enabled by structures that consist of a scientific-actual front, a social-relational front and a cognitive-mental front, mobilize and realize the agency of themselves and of others that can be differentiated as intelligence, imagination and involvement.

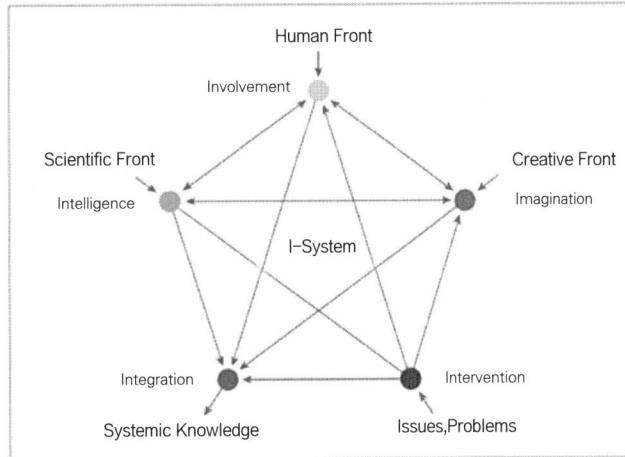

Figure 2-4 I-system: a knowledge construction model

Source: Y.Nakamori, Systems Research and Behavioral Science 23,2006: 3–19.

Y. Nakamori, A.P. Wierzbicki, Z. Zhu, Systems Research and Behavioral Science 28,2011:15–39.

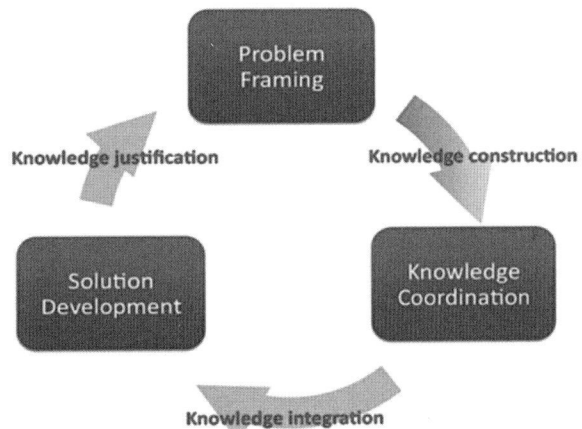

Figure 2-5 Newly Knowledge Science guided approach to decision analysis

Source: Van Nam HUYNH, 2016, Complex Decision Making from the Perspective of Knowledge Science.

I-system provide a systematic and unified view of complex decision problems so as to establish a modeling framework addressing what and how knowledge could be comprehensively constructed for decision analysis, as Figure 2-6. I-system dealing with uncertainty/imprecision, as well as inhomogeneous knowledge, as Figure 2-7.

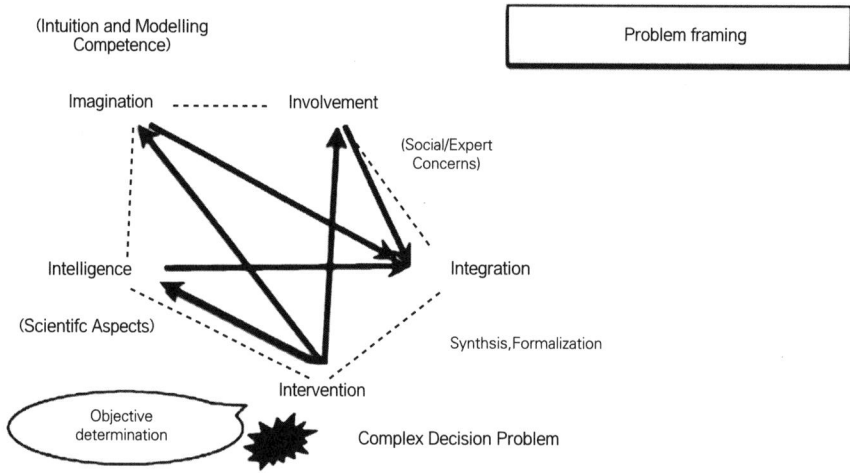

(Intuition and Modelling
Competence)

Problem framing

Imagination ---- Involvement

(Social/Expert
Concerns)

Intelligence — Integration

(Scientifc Aspects)

Synthsis, Formalization

Intervention

Objective
determination

Complex Decision Problem

Figure 2-6 I-system for modeling

Source: Van Nam HUYNH, 2016, Complex Decision Making from the Perspective of Knowledge Science.

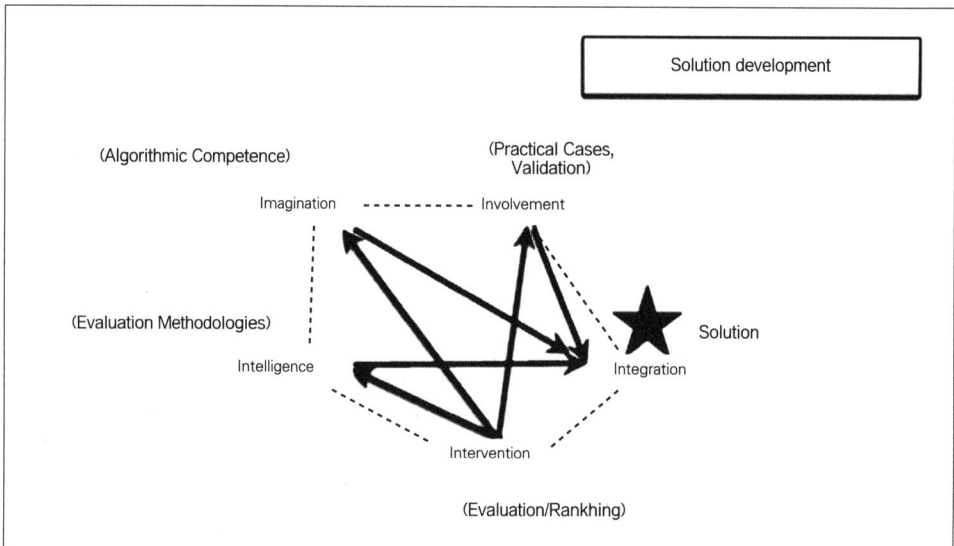

Solution development

(Algorithmic Competence)

(Practical Cases,
Validation)

Imagination ---- Involvement

(Evaluation Methodologies)

Solution

Intelligence

Integration

Intervention

(Evaluation/Rankhing)

Figure 2-7 I-system for knowledge coordination

Source: Van Nam HUYNH, 2016, Complex Decision Making from the Perspective of Knowledge Science.

I-system dealing with uncertainty/imprecision, as well as inhomogeneous knowledge. Integration of synthesized decision knowledge and methodological knowledge in order to develop a sound solution for complex decision problem, as Figure2-8,Figure2-9.

A summary of the theory of knowledge construction systems is given below, which consists of three fundamental parts:

(1)The knowledge construction system: A basic system to collect and synthesize a variety of knowledge, called the I-System, which itself is a systems methodology (Nakamori, 2000/2003).

(2)The structure-agency-action paradigm: A sociological interpretation of the I-system to emphasize the necessary abilities of actors when collecting and synthesizing knowledge (Nakamori and Zhu, 2004).

(3)The evolutionary constructive objectivism: A new episteme to create knowledge and justify collected and synthesized knowledge (Wierzbicki & Nakamori, 2007).

The main characteristics of this theory are as below.

Fusion of the purposiveness paradigm and purposefulness paradigm: With the I-system, we always start by searching for and defining the

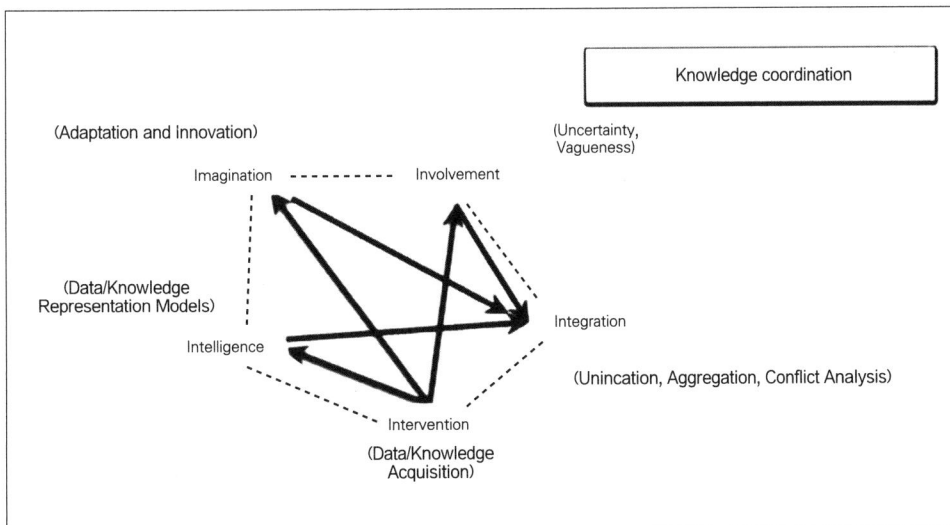

Figure 2-8 I-system for solution development

Source: Van Nam HUYNH, 2016, Complex Decision Making from the Perspective of Knowledge Science.

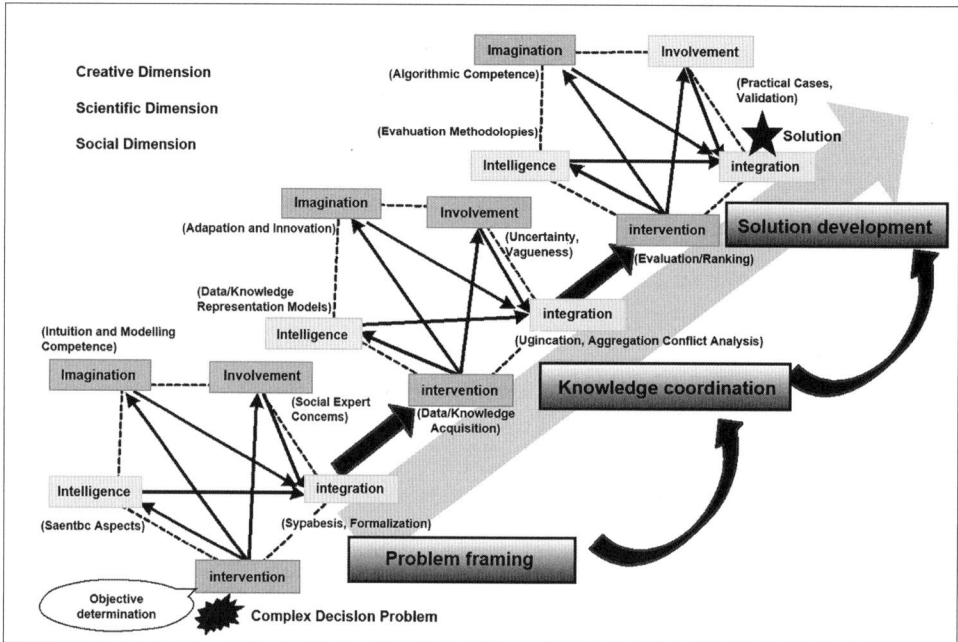

Figure 2-9 An I-system guided-approach to modeling and solution development for complex decision problems

Source: Van Nam HUYNH, 2016, Complex Decision Making from the Perspective of Knowledge Science.

problem according to the purposefulness paradigm. Since the I-system is a spiral-type knowledge construction model, in the second cycle we use the I-system to find solutions according to the purposiveness paradigm. However, it is almost always the case that when we find an approximate solution, we face new problems.

Interaction of explicit knowledge and tacit knowledge: An important idea of Nonaka and Takeuchi (1995) is that new knowledge can be obtained by the interaction between the explicit and the tacit knowledge. The use of the I-system means that we must inevitably deal with objective knowledge such as scientific theories, available technologies, social-economic trends. As well as subjective knowledge such as experience, technical skills, hidden assumptions, paradigms.

Involvement of knowledge coordinators: the theory requires people who accomplish the knowledge synthesis. Such persons need to have the abilities of knowledge workers and innovators in wide-ranging areas. However, they cannot achieve satisfactory results unless they also possess the ability to coordinate the opinions and values of diverse people. An educational system should be established to train human resources who will promote knowledge synthesis in a systemic manner.

2.2

Overview of the Fashion Design Education

2.2.1 Definition of Fashion Design

Fashion design is the art of application of design and aesthetics or natural beauty to clothing and accessories. Fashion design is influenced by cultural and social attitudes, and has varied over time and place. Fashion designers work in a number of ways in designing clothing and accessories such as bracelets and necklaces. Because of the time required to bring a garment onto the market, designers must at times anticipate changes to consumer tastes (Wikipedia, 2011). Fashion design is generally considered to have started in the 19th century with Charles Frederick Worth who was the first designer to have his label sewn into the garments that he created. Before the former draper set up his maison couture (fashion house) in Paris, clothing design and creation was handled by largely anonymous seamstresses, and high fashion descended from that worn at royal courts. Worth's success was such that he was able to dictate to his customers what they should wear, instead of following their lead as earlier dressmakers had done. The term couturier was in fact first created in order to describe him. While all articles of clothing from any time period are studied by academics as costume design, only clothing created after year 1858 is considered as fashion design.

2.2.1.1 Types of Fashion

Different kinds of wears belongs to different types of fashion, as shown in Table 2-1.

(1)Haute couture.

Until the 1950s, fashion clothing was predominately designed and manufactured on a made-to-measure or haute couture basis (French for high-sewing), with each garment being created for a specific client. A couture garment is made to order for an individual customer, and is usually made from high-quality, expensive fabric, sewn with extreme attention to detail and finish, often using time-consuming, hand-executed techniques. Look and fit take priority over the cost of materials and the time it takes to make. Due to the high cost of each garment, haute couture makes little direct profit for the fashion houses, but is important for prestige and publicity.

(2)Ready-to-wear.

Ready-to-wear, clothes are a cross between haute couture and mass market. They are not made for individual customers, but great care is taken in the choice and cut of the fabric. Clothes are made in small quantities to guarantee exclusivity, so they are rather expensive. Ready-to-wear collections are usually presented by fashion houses each season during a period known as Fashion Week. This takes place on a citywide basis and occurs twice a year. The main seasons of Fashion Week include, spring/summer, fall/winter, resort swim and bridal.

(3)Mass market.

Currently the fashion industry relies more on mass-market sales. The mass market caters for a wide range of customers, producing ready-to-wear garments using trends set by the famous names in fashion. They often wait around a season to make sure a style is going to catch on before producing their own versions of the original look. To save money and time, they use cheaper fabrics and simpler production techniques which can easily be done by machine. The end product can therefore be sold much more cheaply.

2.2.1.2 Design Process for Fashion Design

Four examples of Fashion Thinking models are given to show the impact of different approaches.

(1)An example of the first type.

It is fusion style of product development (Hughes, 1995) concentrating

Table 2-1 Area of fashion design

Area	Brief	Market
Women's day wear	Practical, comfortable, fashionable	Haute couture, ready-to wear, mass market
Women's evening wear	Glamorous, sophisticated, suited for the occasion	Haute couture, ready-to-wear, mass market
Women's lingerie	Glamorous, comfortable, washable	Haute couture, ready-to-wear, mass market
Men's day wear	Casual, practical, comfortable	Tailoring, ready-to-wear, mass market
Men's evening wear	Smart, elegant, formal, apt for the occasion	Tailoring, ready-to-wear, mass market
Kids' wear	Trendy or classy, practical, washable, functional	Ready-to-wear, mass market
Girls' wear	Pretty, colorful, practical, washable, inexpensive	Ready-to-wear, mass market
Teenager girl wear	Colorful, comfortable, glamorous, pretty, cute	Ready-to-wear, mass market
Jeans wear	Unisex, democratic, comfortable, practical, functional	Ready-to-wear, mass market
Swimwear	Trendy, stylish, practical, functional, colorful	Haute couture, ready to-wear, mass market
Sports wear	Comfortable, practical, well-ventilated, washable, functional	Ready-to-wear, mass market
Knitwear	Right weight and color for the season	Ready-to-wear, mass market
Outerwear	Stylish, warm, right weight and color for the season	Ready-to-wear, mass market
Bridal wear	Sumptuous, glamorous, classic	Haute couture, ready-to-wear, mass market
Accessories	Striking, fashionable	Haute couture, ready to-wear, mass market
Performance wear	Sporty, dependent on the sport	Ready-to-wear, mass market

on the relationships between the sets of activities and responsibilities of each function involved in the fashion design process (Figure 2-10). Although, the model still demonstrates the task in "sequence structure" to some extent, the cross-functional collaboration is clearly expressed. However, it is unable to describe the key elements and the activities that actually need to be carried out.

(2)An example of the second type.

It is Fashion Design cycle (Rhodes, 1995) focusing on the activities within the process (Figure 2-11). As a result, the fashion thinking model is presented in a linear structure. Although, it is able to address many key elements, e.g. input from all participants and relationships between tasks and disciplines, the linear structure prevents it from capturing the active interaction and influence of each key element.

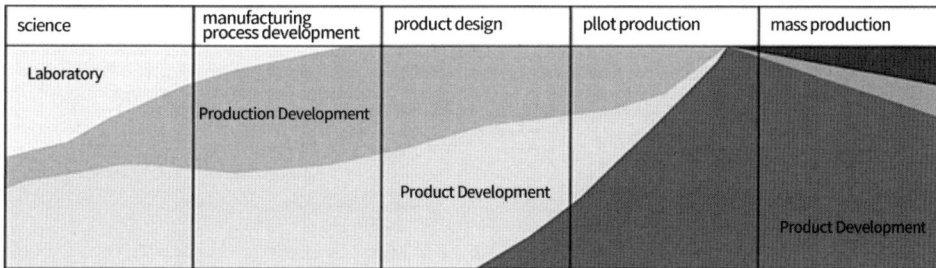

science	manufacturing process development	product design	pllot production	mass production

Figure 2-10 Matsushita industry's fusion style of fashion design process (Hughes, 1995)

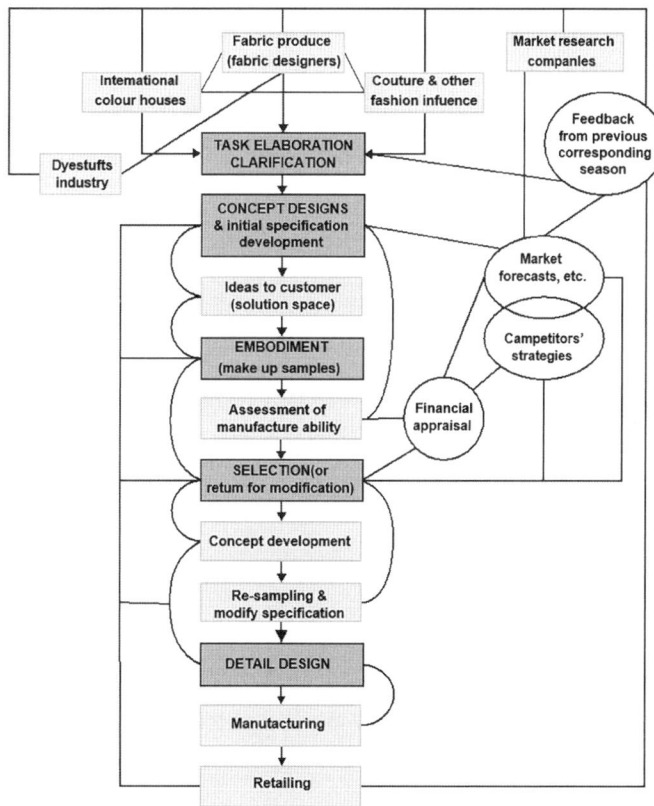

Figure 2-11 Fashion design cycle (Rhodes,1995)

(3)An example of the third type.

It is Product Development framework (Kallal & Lamb, 1993; cited in Le Pechoux, Little, & Istook, 2001) emphasizing the key elements affecting fashion design process, e.g. brand, market response, competition, etc. (Figure 2-12). Therefore, all the elements and their relationship are

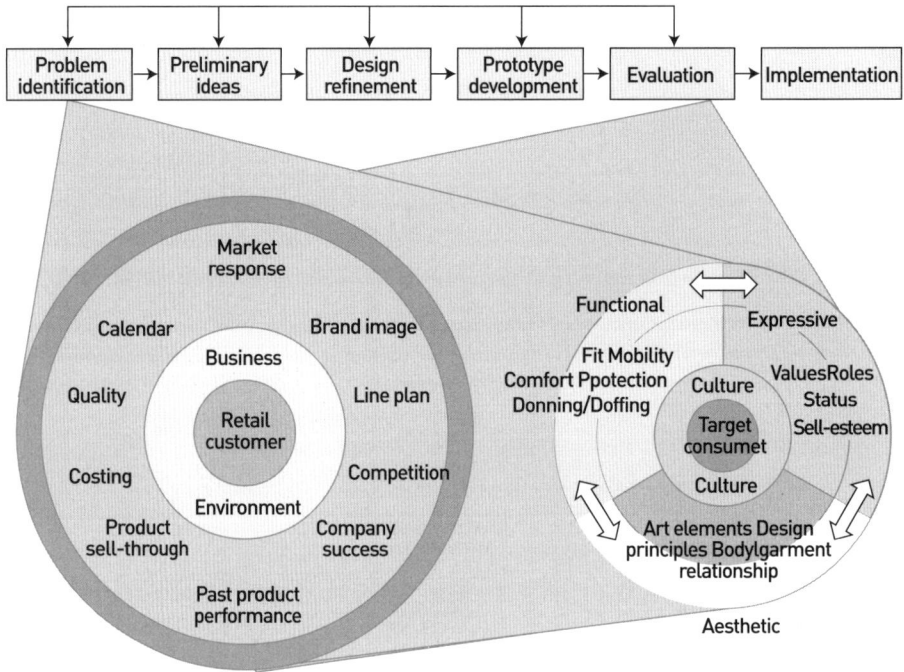

Figure 2-12 Fashion product development framework emphasizing retailer's influence (Kallal and Lamb, 1993;cited in Le Pechoux, Little and Istook, 2001)

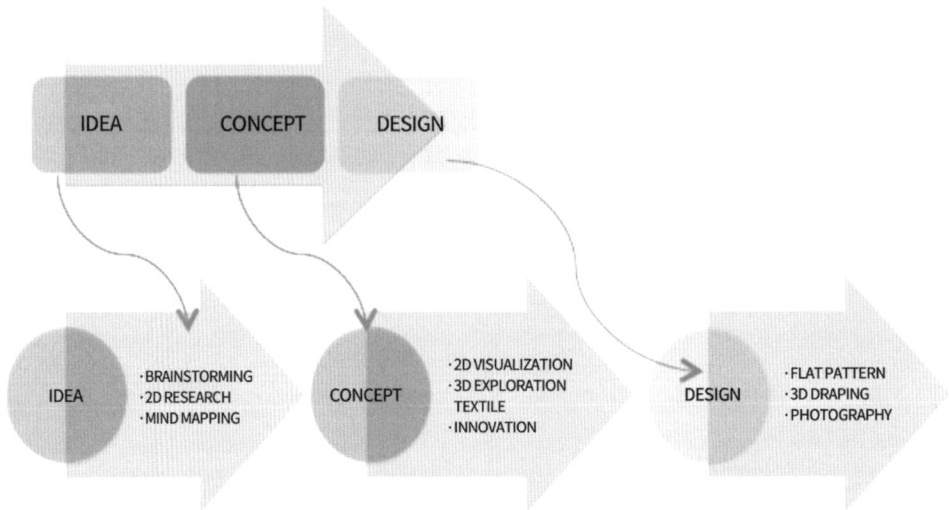

Figure 2-13 Fashion design process system thinking (Simon Seivewright, 2007)

addressed and clearly presented. However, it is unable to demonstrate which discipline is responsible for which aspect.

(4)An example of the fourth type .

It is Fashion Design process system thinking (Simon Seivewright, 2007) emphasizing the key elements affecting fashion design process, e.g. idea, concept, it seeks to understand the creative design process, how research plays a a role in forming the individual design identity and offering a useful window into fashion designers creative thinking. With an investment in ideas at its core, it enables a student to build on their knowledge and practice. The actual doing of craft or thinking through of ideas are all valuable, and from a personal system of designing within the parameters of modern fashion practice (Figure 2-13). However, it is unable to demonstrate which discipline is responsible for which aspect.

2.2.2 The Main Design Thinking Methodologies in Design Education

All forms of professional design education can be assumed to be developing design thinking in students, even if only implicitly, but design thinking is also now explicitly taught in general as well as professional education, across all sectors of education (Archer L. B. ,et al., 1979). Design as a subject was introduced into secondary schools' educational curricula in the UK in the 1970s, gradually replacing and/or developing from some of the traditional art and craft subjects, and increasingly linked with technology studies. This development sparked related research studies in both education and design.

2.2.2.1 Design Thinking

Design thinking refers to creative strategies designers use during the process of designing. It has also been developed as an approach to resolve issues outside of professional design practice, such as in business and social contexts.

The origins of regarding design thinking as a particular approach to creatively solving problems lie in the development of creativity techniques in the 1950s and the development of new design methods in the 1960s. L. Bruce Archer was perhaps the first author to use the term "design thinking' in his book "Systematic Method for Designers" (1965). The notion of design as a "way of thinking" in the sciences can be traced to Herbert A. Simon's 1969 book The Sciences of the Artificial, and in design engineering to Robert McKim's 1973 book Experiences in Visual Thinking. Bryan Lawson's 1980 book How Designers Think, primarily addressing design in architecture, began a process of generalising the concept of design thinking. A 1982 article by Nigel Cross established some of the intrinsic qualities and abilities of design thinking that made it relevant in general education and thus for wider audiences. Peter Rowe's 1987 book Design Thinking, which described methods and approaches used by architects and urban planners, was a significant early usage of the term in the design research literature. Rolf Faste expanded on McKim's work at Stanford University in the 1980s and 1990s, teaching "design thinking as a method of creative action." Design thinking was adapted for business purposes by Faste's Stanford colleague David M. Kelley, who founded the design consultancy IDEO in 1991. Richard Buchanan's 1992 article "Wicked Problems in Design Thinking" expressed a broader view of design thinking as addressing intractable human concerns through design.

Thinking like a designer can transform the way organizations develop products, services, processes, and strategy. This approach, which IDEO calls design thinking, brings together what is desirable from a human point of view with what is technologically feasible and economically viable. It also allows people who aren't trained as designers to use creative tools to address a vast range of challenges.

As an approach, design thinking taps into innate human capacities that are overlooked by more conventional problem-solving practices (Brown T & Wyatt J, 2010). The process is best thought of as a system of overlapping spaces rather than a sequence of orderly steps: inspiration, ideation, and implementation (Brown T & 2008); or alternatively: empathize, define, ideate, prototype and test (Figure 2-14). Projects may loop back through inspiration, ideation, and implementation more than once as the team refines its ideas and explores new directions. Therefore, design thinking can feel chaotic, but over the life of a project, participants come to see that the process makes sense and achieves results, even though its form differs from the linear, milestone-based processes that organizations typically undertake (Brown T & Wyatt J, 2010).

(1)Empathize.

Empathize or empathy is the foundation of Design Thinking. As a Design Thinker we should be able to understand all the feelings, thoughts, complaints, expectations and habits of the people whom we'll make the design ideas and solutions. Therefore every sense, feeling and thought

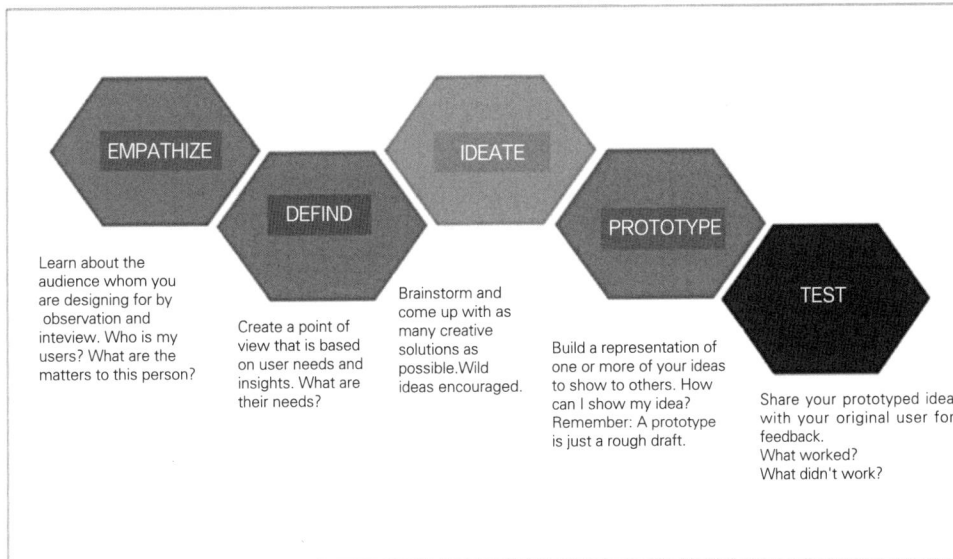

Figure 2-14 Design thinking modes (Stanford Design School & IDEO)

should be focused and centered to the person, supposing the English term "Putting ourselves into their shoes". In this stage we can ask anything, to better understand.

How do we empathize? Easy things we can do are three things:

①Observe (observed); observe the behavior and habits of the person as what he/she is in his/her environment. Watch how he interacts with the surrounding and how he uses any goods or services that are available to him. We quite simply by paying attention without having to interact with him, we can record it with a camera.

②Engage (involved); here we can interact by asking or interviewing the person. Previously we have prepared a series of questions with a sense as courious as we may to ask what is done by the person and what he wants. The key is to ask in depth, "Why? Why why?".

③Immerse (into the field); here we look at the immediate context and the environment in which people live or work, and how they interact with their environment. Usually we go to her workplace, or her home or a place she does something. Just like Ethnographic studies, Immersion is also doing the same thing. Go directly to the context of the environment makes us understand why she behaved as she did.

(2)Define.

Results from empathize mode becomes material for us to define any findings of observation, engagement and field immersion. In this mode we pay attention to every detail of the data and information we found. Then we focus again to the insights, needs, and scope of the challenges faced by the person. Results of this mode is a form of problem statement or formulation of the problems and challenges faced by the person and also what is the scope of the design space of innovation which we will do. In "Define", we formulate the problem statement as the focus of the problems faced by the person, from our point of view.

(3)Ideate.

If the problem statement has been formulated with a fitting and focus, then the next step we generate a wide range of ideas to address the challenges and meet the needs of the person. In this mode we can think of flaring to the creative process that is as large as possible. Surely, the creative ideas that we can think of to be tune in with the problem statement in the previous "Define" mode. In addition, it is also important to be able to spawn ideas that are unique and original.

(4)Prototype.

Well next, after all sorts of ideas have been thought and recorded properly, it is time to translate these ideas in a more obvious physical or visualization. This is where we make the prototype phase or prototyping. In this mode we focus on ideas which are most likely and best for a prototype we make. Prototype can be in any form, from simple things like images on a sheet of paper, or sketch-style building architect, or up to a more advanced prototype of such a program or computer application. So the prototype is not necessarily good or wrong immediately. The important thing with the prototype can describe the idea that we want and make everyone able to interact with our idea of it.

(5)Testing.

Prepare our prototype and then try to do a series of tests to the target user we have visited or interviewed before. Try to see how people interact with our prototype. Look closely, whether the features that we've designed, well received by the person. Or are there things that are or are not in accordance with our expectations when the person using our prototype? Here we note what are the points to increase our prototype so it could be better.

2.2.2.2 User Experience Design (UX/UXD)

User experience Design (abbreviated as UX/UXD) is the process of enhancing customer satisfaction and loyalty by improving the usability, ease of use, and pleasure provided in the interaction between the customer and the product. The field of user experience design is a conceptual design discipline and has its roots in human factors and ergonomics, a field that, since the late 1940s, has focused on the interaction between human users, machines, and the contextual environments to design systems that address the user's experience (Karen, 2016) With the proliferation of workplace computers in the early 1990s, user experience started to become a concern for designers.

It was Donald Norman, a user experience architect, who coined the term "user experience", and brought it to a wider audience. I invented the term because I thought human interface and usability were too narrow. I wanted to cover all aspects of the person's experience with the system including industrial design graphics, the interface, the physical interaction and the manual. Since then the term has spread widely, so much so that it is starting to lose its meaning. —Donald Norman (Merholz & Peter, 2007)

User experience design includes elements of interaction design, information architecture, user research, and other disciplines, and is concerned with all facts of the overall experience delivered to users (Figure 2-15). Following is a short analysis of its constituent parts:

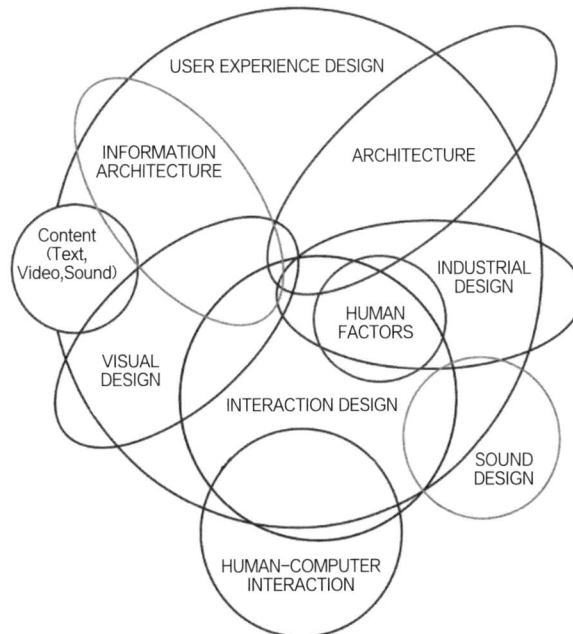

Figure 2-15 User experience design diagram

(1)Visual design.

Visual design, also commonly known as graphic design, user interface design, communication design, and visual communication, represents the aesthetics or look-and-feel of the front end of any user interface. Graphic treatment of interface elements is often perceived as the visual design. The purpose of visual design is to use visual elements like colors, images, and symbols to convey a message to its audience. Fundamentals of Gestalt psychology and visual perception give a cognitive perspective on how to create effective visual communication.

(2)Information architecture.

Information architecture is the art and science of structuring and organizing the information in products and services to support usability and findability.

In the context of information architecture, information is separate from both knowledge and data, and lies nebulously between them. It is information about objects (Garrett & Jesse, 2011). The objects can range from websites, to software applications, to images et al. It is also concerned with metadata: terms used to describe and represent content objects such as documents, people, process, and organizations.

(3)Navigation design.

Navigation design is the way in which the interface elements are placed so as to regulate the users movement through the information architecture and make it simple (Jesse, 2017). Structuring, organization, and labeling: Structuring is reducing information to its basic building units and then relating them to each other. Organization involves grouping these units in a distinctive and meaningful manner. Labeling means using appropriate wording to support easy navigation and findability.

(4)Finding and managing.

Find-ability is the most critical success factor for information architecture. If users are not able to find required information without browsing, searching or asking, then the find-ability of the information architecture fails. Navigation needs to be clearly conveyed to ease finding of the contents.

(5)Interaction design.

There are many key factors to understanding interaction design and how it can enable a pleasurable end user experience. It is well recognized that building great user experience requires interaction design to play a pivotal role in helping define what works best for the users. High demand for improved user experiences and strong focus on the end-users have made interaction designers critical in conceptualizing design that matches user expectations and standards of the latest UI patterns and components. While working, interaction designers take several things in consideration. A few of them are (Psomas & Steve ,2007):

①Defining interaction patterns best suited in the context.

②Incorporating user needs collected during user research into the designs.

③Features and information that are important to the users.

④Interface behavior like drag-drop, selections, and mouse-over actions.

⑤Effectively communicating strengths of the system.

⑥Making the interface intuitive by building affordances.

⑦Maintaining consistency throughout the system.

In the last few years, the role of interaction designer has shifted from being just focused on specifying UI components and communicating them to the engineers to a situation now where designers have more freedom to design contextual interfaces which are based on helping meet the user needs (Lowgren, Jonas, 2015). Therefore, User Experience Design evolved into a multidisciplinary design branch that involves multiple technical aspects from motion graphics design and animation to programming.

2.3
Review of Relevant Knowledge Areas

2.3.1 Overview of Smart Clothing Design
2.3.1.1 Definition of Smart Clothing

Clothing is an environment that we need to wear every day. Clothing is special because it is personal, comfortable, close to the body, and used almost anywhere at any time (Kirstein, et al., 2005). People enjoy clothing, with pleasures associated with its selection and wearing. There is a need for an "ambient intelligence" in which intelligent devices are integrated into the everyday surroundings and provide diverse services to everyone. As our lives become more complex, people want "ambient intelligence" to be personalized, embedded, unobtrusive, and usable any time and anywhere. Clothing would be an ideal place for intelligent systems because clothing could enhance "our capabilities without requiring any conscious thought or effort" (Mann, 1996). Clothing can build a very intimate form between human – machine interaction.

Smart clothing is a "smart system" capable of sensing and communicating with environmental and the wearer's conditions and stimuli. Stimuli and responses can be in electrical, thermal, mechanical, chemical, magnetic, or other forms (Tao, 2001). Smart clothing differs from wearable computing in that smart clothing emphasizes the importance of clothing while it possesses sensing and communication capabilities (Barfield, et al., 2001). Wearable computers use conventional technology to connect available electronics and attach them to clothing. The functional components are still bulky and rigid portable machines and remain as non-textile materials. While constant efforts have been made toward miniaturization of electronic components for wearable electronics, true "smart clothing" requires full textile materials for all components. People prefer to wear textiles since they are more flexible, comfortable, lightweight, robust and washable (Kirstein, et al., 2005). To be a comfortable part of the clothing, it is necessary to embed electronic functions in textiles so that both electronic functionality and textile characteristics are retained. Smart clothing should be easy to maintain and use, and washable like ordinary textiles. Therefore, combining wearable technology and clothing/textile science is essential to achieve smart clothing for real wearability.

Smart clothing formed from the functional expansion on the traditional fabrics and clothing by virtue of computers and materials science and technology, has evolved from the traditional clothes added with sensors and other electronic devices and embedded with smart chips, aiming to make such clothes have the function of computers. Then the smart clothing can produce a variety of "smart" responses to the variations in the human body and current environment so as to effectively assist people working in a particular environment. For instance, it will be applied to adjust the temperature automatically in accordance with the current environment and to detect dangerous environment in order to avoid injuries, in the meanwhile, to monitor people's health and physical conditions and so on. Besides, the research topic of smart clothing belongs to the interdisciplinary studies covering computer science, material science and clothing design (Figure 2-16). In such a field of investigation, on the one hand, the computer and material scientists focus on the effective integration of sensors and other electronic components into the fabrics, so that the fabrics can effectively sense the variations of temperature, pressure, sound and other parameters in the surrounding environment. On the other hand, the apparel designers are required to design the apparels that suits a particular environment and customers, depending on the specific fabrics and the roles of apparels.

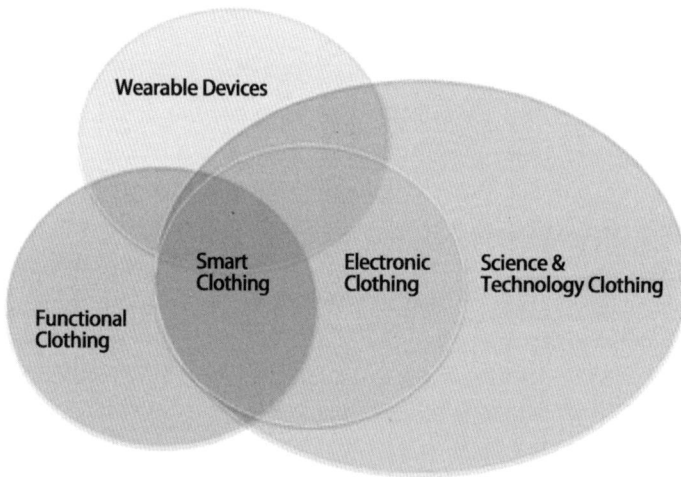

Figure 2-16 Smart clothing relationship with different area

Source: Qiu Chunyan, Hu Yue, 2015.The Review of Smart Clothing Design Research Based on the Concept 3F+1.

2.3.1.2 Evolution History of Smart Clothing

Product scenarios of these projects were compared and analysed in order to discover the ideas and inspirations behind the developments. Moreover, they reveal each team's vision of the future and how they respond to new possibilities that Smart Clothing brings. The analytical results indicate that the design evolution can be divided into three periods (Figure 2-17):

(1)In the first period, 1980 to 1997.

The design approach was regarded as technology-driven, since most research and developments focused on Wearable Computing and applications of advanced technologies.

(2)In the second period, 1998 to 2000.

The awareness and involvement of the fashion and textile sector significantly rose. Consequently, the number of the collaborative projects between electronic and fashion fields rapidly increased.

(3)In the third period, 2001 to 2004.

The number of research studies and Smart Clothing applications available in the market increased dramatically. As multidisciplinary approach and user-centred design are widely adopted by most development teams, the development process became more complex. The boundary of Smart Clothing applications expanded into new areas.

GapKid's Hoodio (a sweatshirt with machine-washable FM radio embedded) was launched in

November 2004 Adidas introduced "Adidas 1" – self-adapting shoes in May 2004

i-Wear Fashion Show in Paris (05/11/03) featured Alexandra Fede's high tech collections Nokia

presented the new wearable range in 2003 and planned to launched the products in 1st quarter of 2004

Tokyo University's Transparent Clothes was published on BBC News website in February 2003

Vivometrics's Life Shirt, which monitored, recorded and analysed physiologic data, were published and launched

Fede's massage dress was presented at Avantax International Innovation Forum and Symposium in May 2002

KSI's Smart Wear was first employed at German Championship in Athletics

Cornell University's Smart Jacket was presented in 1st ICEWES's Conference in 2002 | Researcher pursues PhD

Infineon Technology and Master School of Fashion (Munich) presented Wearable Electronics range in April 2002 | Launched product (2004)

Central St. Martins College of Art and Design presented Clothing Contra Crime project in June 2001

The North Face launched METS or a self-heating jacket

Lunar Design won the award for "BLU" jacket concept in March 2001 | **3rd wave**

Panasonic and Polo Jean Co launched Tech Style collection in Fall 2001

Mmode Group started in July 2001 and presented Smart Materials Research in September 2001

Pioneer Corporation and fashion companies started Media Fashion Project in March/April 2001 | Presented new results in March 2002

Motorola Inc. presented wearable concepts called Smart Communication™ | Worked with Frog Design (2003)

ICEWES set up High-Tech Fashion Network | Had 1st conference in December 2002

CTIA's fashion show 'Fashion in Motion' started in 2001 | Latest show (22-24/03/2004)

IEE Eurowear conference first started in 2001 – 'Wear me' exhibition (4-5/09/2003)

2nd wave | France Telecom presented its first functional prototype in May 2000 | Presented more concepts in 2002 and 2003

Enlighted Design Inc presented Illuminated Clothing in April-May 2000

Starlab's i-Wear Intelligent Clothing Consortium started in 1999

Central St. Martins College of Art and Design's started MA Textile Future | The course have been running since 1999

MIT Media Lab, IDEO and BMW started project together in September 1999

Charmed Technology started Brave New Unwire World Fashion Show | Fashion Show was organised again in 2000

VectraSence (spin-off company from MIT) was founded in 1999 | Presented and sold ThinkShoe since 2001

ElekTex™ or Eleksen found in September 1998

Brunel University's Sensory Fabric project published in July 1998

Reima Tutta set up Clothing+ and started Smart Clothing projects | Presented concepts (2000) and launched wearable products since 2001

IEEE's ISWC conference first started in 1997 | 8th annual conference was 2-5/11/04

Bristol University published Wearable Computing project | Presented at IEE Eurowear in 2003

MIT Media Lab started Fabric Computing Interface project | Presented and published in 2000 and set up IFM in 2001

Philips and Levi's started Wearable Electronics project | Unveiled in August 1999 and sold first ICD+ in September 2000

MIT Media Lab organised Fashion Show (15/10/97)

Sensatex and US military started the project | Launched Smart Shirt in 1st quarter of 2001

MIT Media Lab started MIThril project (1996) | Published (1997)

Philips started wearable electronics (1995)

MIT Media Lab started "Lizzy" project (1992/3) | Worked with Nike (1999) and present new medical application (2003)

Prof. Mann started Cyberman project | **1st wave**

| | 1980 | 1996 | 1997 | 1998 | 1999 | 2000 | 2001 | 2002 | 2003 | 2004 |

Figure 2-17 Diagram explaining evolution of the smart clothing development
Source: Wearable Technology-Market Assessment. 2015. 17 Apr. 2016.

With the growing interest in smart clothing from the industry as well as academia, we anticipate that the area of smart clothing will continue to expand. Growth is expected to occur in applications requiring dedicated functions and in everyday clothing in which the wearer's emotions could be recognized and expressed.

Figure 2-18 summarizes the components of smart clothing technology, the services that smart clothing can provide, and examples of the applications. Individual components of a smart clothing system, i.e., interface, communication, data management, energy management, and integrated circuits, are combined and work together to form services such as information, communication, assistance, aesthetic, affective, etc. An example of information application is a jacket with a GPS, providing positioning and location data. Communication applications include a jacket in which a mobile phone is integrated. An example of affective smart clothing is the Sensor Sleeve, which detects affective gestures and enables emotional messages to be exchanged remotely between people. Smart clothing is likely to expand from a function oriented system to a system that focuses not only on the function but also affective states of the wearer. We are living in an era where human beings search for social interaction and relationships and look for ways to communicate with others and express our emotions. In the future,

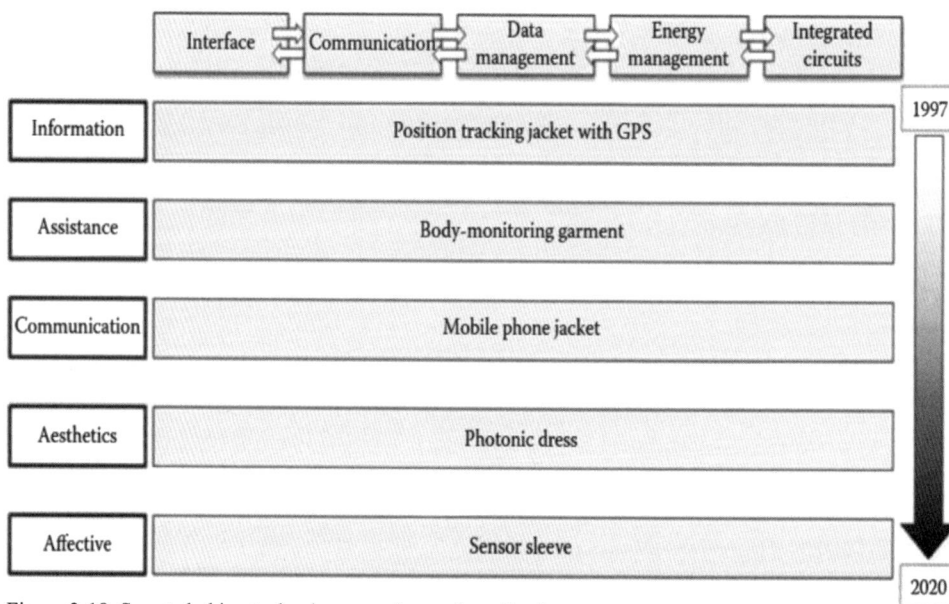

Figure 2-18 Smart clothing technology, service, and applications

Source: Gilsoo Cho, Seungsin Lee, Review and Reappraisal of Smart Clothing, © 2010 by Taylor and Francis Group, LLC.

smart clothing may sense the wearer's feelings and respond to emotions by changing its shape, color, scent, and so forth.

2.3.2 Design Process for Smart Clothing

2.3.2.1 Design Needs for Smart Clothing

In all the fields of industrial design, "designing" means a process which starts from an analysis of design needs and ends at their synthesis into some visualized forms, where "design needs" imply expectations held by consumers and manufacturers toward the object to be designed in the contexts of aesthetics, functionality, ergonomics, safety and price. In short, design needs to indicate what should be harmoniously reflected in every single "designed" product as conceptually fundamental composites. In other words, it means that critical changes occur in the design process when some new design needs emerge. In the case of smart clothing, the territory of design needs for apparel should be widened to include new categories, such as embedding of digital functions or the interaction between parts of digital function and human body.

In the field of smart clothing, some researchers have analyzed design needs (or requirements) according to their varying definitions. Defining smart clothing as "wearable motherboard" in her analysis on "GTWM" (e.g. "Wearable Mother board" of Georgia Tech), Tao (2001) has categorized the requirements into functionality, connect ability, durability, maintainability, usability in combat, manufacturability, wearability, and affordability. Viewing smart clothing as a garment system of wearable technology, Dunne, Ashdown, and McDonald (2002) have analyzed the needs into the matters of thermal management, moisture management, mobility, flexibility, sizing and fit, durability and garment care. Cho (2004) has defined smart clothing as digital clothing and categorized the requirements into durability, easy care, comfort, safety and aesthetic satisfaction.

Based on previous examinations of design needs, we suggest a total of 10 key categories in design needs for smart clothing: functionality, usability, comfort, maintainability, easy garment care, manufacturability, safety, wearability, durability and appearance. Comfort, wearability, easy garment care and appearance are the matters relevant to the definitions of "a type of clothing", whereas functionality and maintainability are related to the definition of "clothing with digital/mechanical function." Durability and manufacturability are the design needs associated with both definitions. Usability and safety are related to clothing-typed human machine interface.

2.3.2.2 Design Process for Smart Clothing

Viewed as a mental process, a design process involves a set of highly organized procedures of problem solving where various types of information are collected and synthesized into a consistent concept and finally transferred into a visual form. As every case of problem-solving process does, design process varies along with the feature and structure of design issues to be solved. Smart clothing is a good example of where such an innovation in the design process should occur, because it is based on a set of design needs that combine the design needs for clothing with the design needs for digital products. Smart clothing as a product made by integrating digital devices with clothing is different from general clothes in its components, and therefore it requires adoption of the IT (information technology) product development process for its design. Under this assumption, Lee (2006) has developed a smart clothing design process after organizing a tentative design process model for smart clothing and testing the efficiency of this model, applying the general theories for design process to her own experience in case studies.

After considering the general theories for design process, the traditional clothing design process theory (et al., 1998), the sports wear design process theory (Kim, 1999), and the product design process (Ulrich, 2004), Jane McCann (2005), looks at a comparatively new and unique design discipline that has been given little prior consideration. The concept of "wearables" crosses the boundaries between many disciplines. A gap exists for a common "language" that facilitates creativity and a systematic design process. Designers of smart clothing require guidance in their enquiry, as gaining an understanding of issues such as; usability, manufacture, fashion, consumer culture, recycling and end user needs can seem forbidding and difficult to priorities. The representation of the "critical path" is intended as a tool to guide the design research and development process in the application of smart technologies (Figure 2-19).

Lee (2006) first reviewed and compared the apparel design process with that of other commercial products, matching the similar process steps. A smart clothing design process model was derived by substituting the result of actual case studies into appropriate place of the workflow according to

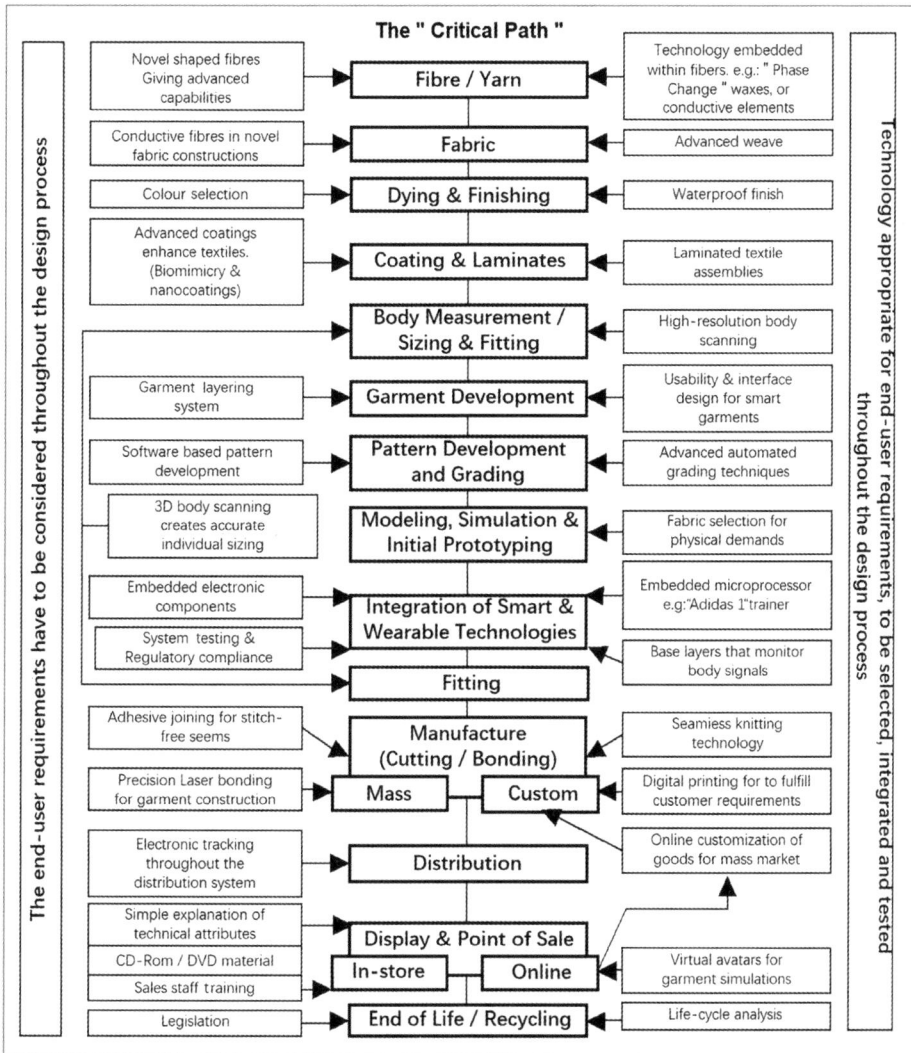

Figure 2-19 The "Critical Path" to be considered during the design process
Source: Jane McCann, 2005,A Design Process for the Development of Innovative Smart Clothing that
Addresses End-User Needs from Technical, Functional, Aesthetic and Cultural View Points.
The 2005 Ninth IEEE International Symposium on Wearable Computers.

time order. It was found that the process smart clothing design contains
a stage that corresponds to the "system level design" of the industrial
product design process. This is the stage that sets up a design plan about

an experimental prototype organized in the previous stage, finds the related subsystem and interface, and modifies the prototype design. In the traditional clothing design process, this stage is almost always omitted (Figure 2-20).

2.3.2.3 Design Process Model of Smart Clothing

Lee (2006) has actually produced the smart photonics clothing by deriving a tentative model of smart clothing design process passing through a series of connected processes as shown in Figure 2-21, and applying the tentative

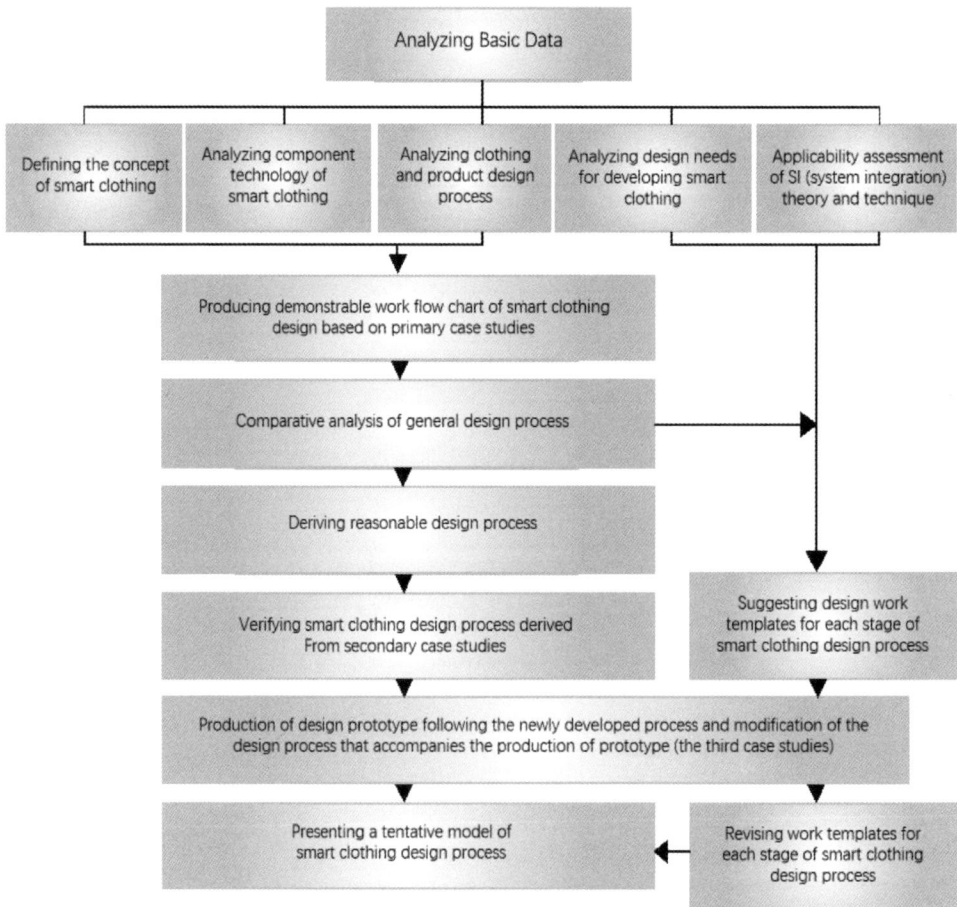

Figure 2-20 Flow chart of development of smart clothing design process model
Source: Lee, Y. 2006. A model of design process for digital-color clothing. Ph.D. dissertation, Yonsei University.

model. The final plan of the smart clothing design process was drawn by analyzing the development process after complementing and revising it. The work template for various types of design to support every step in the design process was developed first. It was then offered to the designers so that they could execute the whole process of smart clothing design by simply following what the work template requested of them.

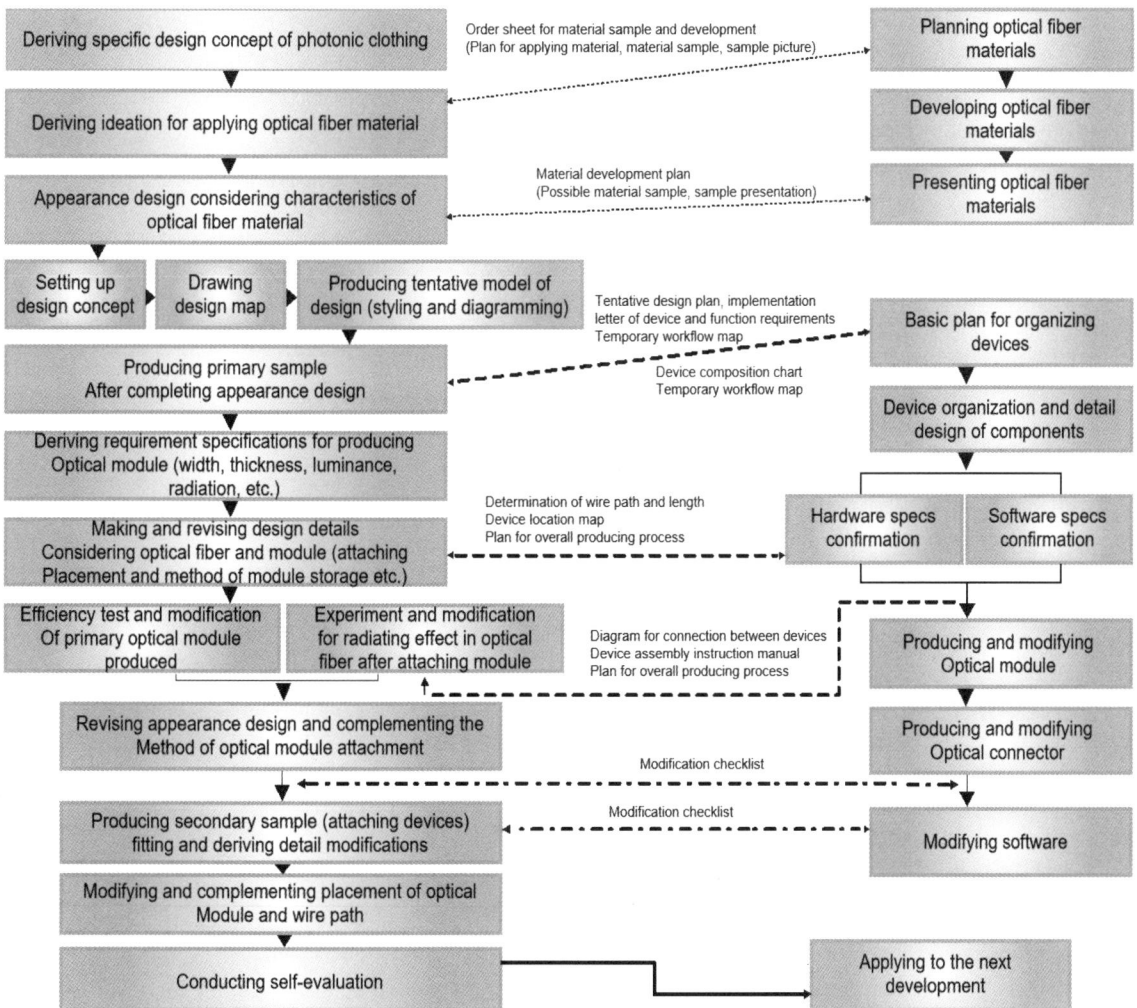

Figure 2-21 Design process model of smart clothing
Source: Lee, Y. 2006. Ph.D. dissertation, Yonsei University.

2.3.3 Applications of Smart Clothing

Smart clothing may serve in various fields since it offers functions for information, assistance, communication, aesthetics, etc. Some products are on the market, but generally developments are in a starting phase; their potential is enormous. Although the study of smart clothing is still in its infancy, currently it has been widely applied in many fields, mainly including health surveillance, entertainment, behavior perception and so on. This thesis will investigate the relevant issues.

2.3.3.1 Health Monitoring

Currently as the most direct and widely applied area of smart clothing, the health surveillance shows a great demand in the field of healthcare. A variety of medical devices in the hospital have been able to accurately measure a large number of human parameters so far, such as body temperature, blood pressure, heart rate, blood glucose, etc. However, such devices are usually limited to be utilized in wards and hospitals, but can't be used to monitor the health of human body in real time in their daily lives. On the other hand, if these devices are skillfully embedded into the clothes of daily life by virtue of the computer hardware and software technology, the persons will be able to obtain relevant health parameters at any time, and then to find the potential diseases in time. This will be beneficial to greatly improve their quality of life. What's more, the smart clothing designed to monitor body temperature and blood pressure has been developed. Such type of smart clothing is embedded with temperature and pressure sensors as well as the single-chip. It's available to collect the data of body temperature and blood pressure in real. time, and then carry out analysis on those data which can be transmitted to the computer by means of the single-chip. The clothes in Figure 2-22 designed by (Dunne et al., 2010) were smart clothing available to monitor the breathing on the basis of pressure sensors, the location of which was marked. Rai et al. (2012) developed clothing based on single-chip and fabric sensors aiming to monitor human health.A. Lymberis et al. (2003) summarized the research and development of smart clothing of monitoring in recent years and predicted the future development trends. Nonetheless, there is still a great challenge to smart clothing of monitoring, and a number of the sensors aimed to measure body parameters are heavy medical machinery that are hard to carry around. In addition, how to embed metal sensors in

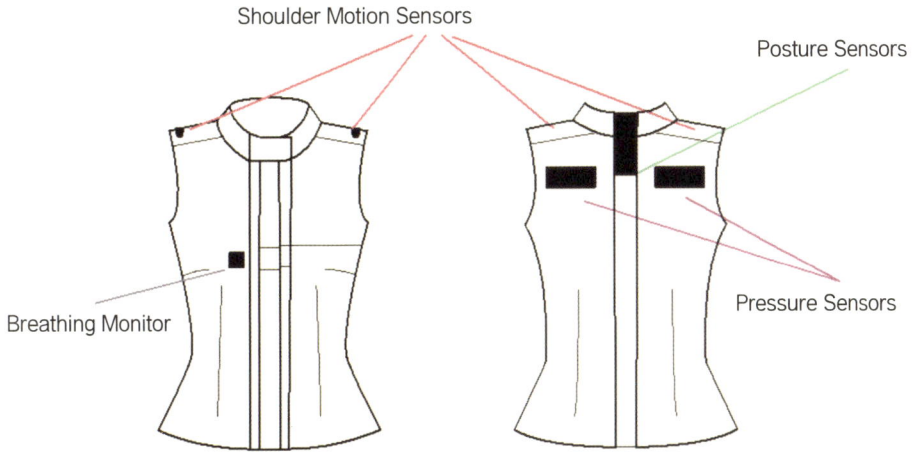

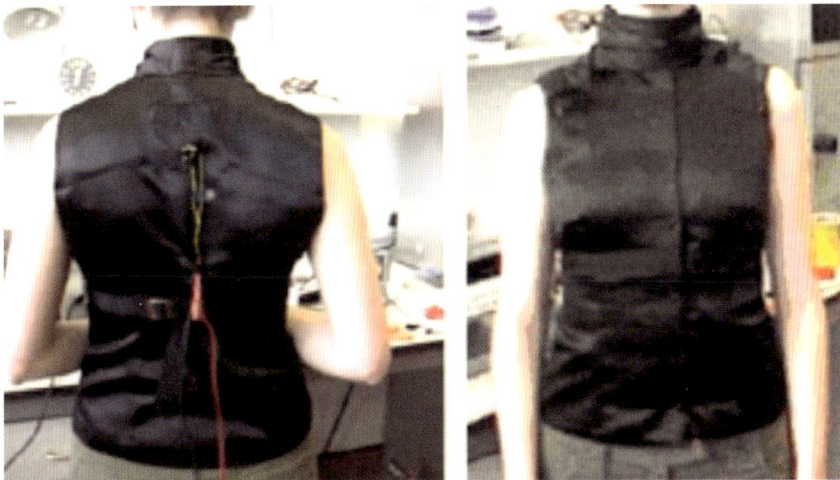

Figure 2-22 An example of smart health monitoring clothes-pressure sensors based breath monitoring

the comfortable clothes is also a challenge for scientists in different fields. However, with the continuous emergence of lightweight sensor materials (such as fabric sensors) and the development of computer intelligent technology, this dream is also getting closer and closer to people's daily lives.

2.3.3.2 Entertainment

As smart clothing can respond to changes in the human body in a variety of interesting ways, it has also been widely applied in the field of

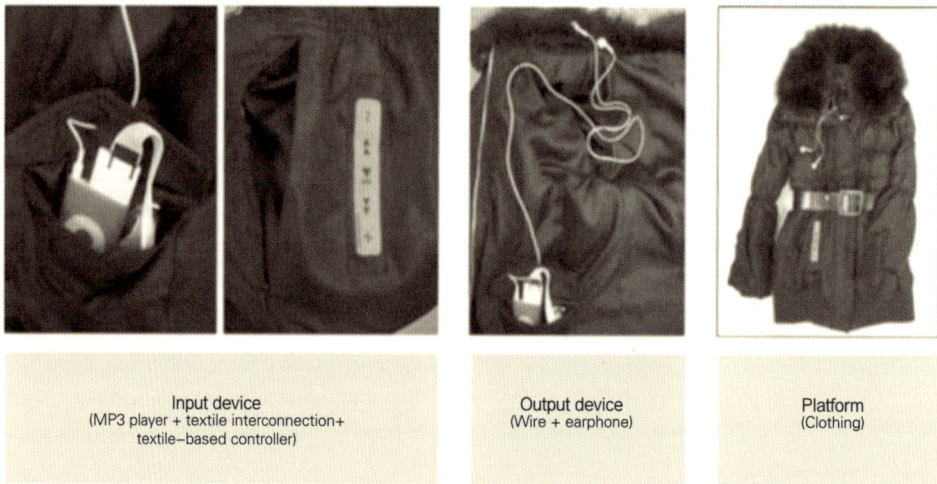

Figure 2-23 Components and clothing design of a MP3-playable jacket "W. α." (Courtesy of Beaucre Merchandizing, Inc.)

entertainment. The Figure 2-23 shows a MP3-playable jacket developed by Seoul National University in Korea, while the Figure 2-24 displays a T-shirt that enables calculation. All of the input devices are made of smart fabrics rather than conventional electronic components. With the further development of sensor technology and computer-based artificial intelligence, smart clothing will definitely be more widely applied in the field of entertainment.

2.3.3.3 Behavior Perception

Behavior perception as the most advanced application of smart clothing, aims to have the clothing with the characteristics of "the second skin of the body". It not only perceives the body's activities, postures and emotions, but also simulates human hearing, smell and touch, and even receives the signals the body can't be perceived. Since the human body perceives a limited range of signals, the computers can be utilized to enhance the body's perception of the surrounding environment. The Figure 2-25 depicts an olfactory skirt developed by the research team of Dr. Jenny Tillotson (2009) at the University of the Arts London in Britain. He can judge the state of human

Figure 2-24 Pressure sensors based calculator embedded into T-shirt

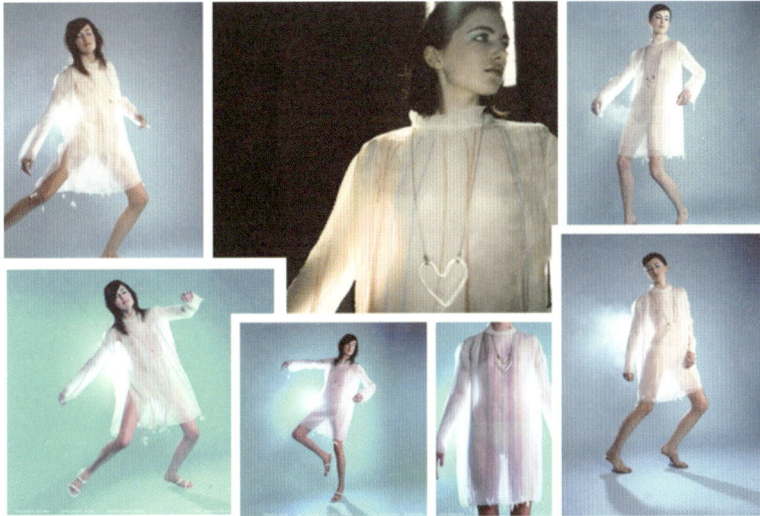

Figure 2-25 A smart dress that can smell

Figure 2-26 Bubelle emotion sensing dress

body such as fatigue, tension, excitement, or even analyze whether someone has the feeling of love, in line with the smell of the human body. This is based on the fact that everyone has their own unique odor like fingerprints, which has been scientifically proven. The Figure 2-26 is "Bubelle emotion sensing dress". This prototype by Philips Design gives us a look into the future of fashion where clothes does not merely to protect, but also reflect our emotions making it a forward form of communication. The first layer of the dress contains biometric sensors which projects emotion which comes in the form of colorful lights onto the second layer, the outer textile. It is both stunning and functional.

Besides, one of the most significant applications of this smart clothing is to assist the daily life of the handicapped. For instance, the vision sensors and computer-based vision technology embedded in clothing are utilized to assist blind persons, and the speech recognition and speech generation technology contained in clothing are also helpful for supporting the deaf-mutes. Furthermore, the brain wave sensors and biosignal sensors inserted in clothing can even acquire the ideas from people's minds and then express them. Therefore, the development of this technology will play a very positive role in promoting social development and progress. Although the above applications are heavily dependent on the sensors and computer-based intelligent technology, the clothing design will inevitably play an increasingly significant role with the gradual popularization of such apparels. What's more, various types of electronic devices need to frequently interact with the human body in smart clothing with behavior perception, hence the relevant requirements will be increasingly high for the safety, comfort, aesthetics and other performance of the clothing design.

The widespread application of the computer-embedded system displays a great role in promoting the research of smart fabrics. In essence, the smart clothing can be regarded as a special kind of computer embedded in the traditional fabrics, and it can sense, store data and calculate. In recent years, the research pertinent to "wearable computer" (S.Mann,1997) has become a new topic in the area of the computer-embedded. This topic focuses on how to integrate sensors into clothing and how to accomplish the specific interaction. The concept was firstly proposed by Edward O. Thhorp in 1961 in an attempt to capture and predict the site data in real time in roulette gambling games. He and Claude Shannon co-developed the first wearable computer, a 4-button simulator with the size of cigarette-box and equipped with a data collector. After that, various types of wearable computers were gradually invented. In 1991, the Carnegie Mellon University (CMU) in the USA developed a wearable computer VuMen1 for maintenance, which consisted of a tiny computer and display of glasses-type. In 1994, Steve Mann implemented the wireless Internet access for wearable computers with real-time image (S.Mann,1997). Moreover, Kevin Warwick developed a necklace which was capable of displaying the signals of human nervous system that could be measured by virtue of the electrode array, in the research project in 2002. The necklace could be varied with different colors in the light of different biological signals (Warwic,2004). In comparison, the fastest growing areas were military applications. The U.S. Defense Department has continuously put forth different plans for development, and provided a large number of practical military wearable computers for the construction of the digital army. Currently the troops have been equipped with wearable computers. In recent years, the science and technology community, business community and application industries seem to have seen its application prospects at the same time. Their cooperation can enhance the development of wearable computing technology into a completely new stage. The development of wearable computers has paved the way for smart clothing, however, there is still a long way from people's expectations. The clothing, much different from the rigid computers, usually appears comfortable, soft, aesthetical and safe. The computers cleverly embed into the soft, aesthetical and comfortable fabrics, are likely to achieve the perfect combination of clothing and computers. At present, smart clothing has been applied to a number of areas, including health monitoring (Rai,2012; Lymberis,2003; Malins,2012), guarding the people working in the extreme environments(Rantanen, 2000;L.Van,2004) , assisting the daily lives of the disabled persons or entertainment. And a series of prominent achievements have been made. Although many researchers think that smart clothing will become one of the major trends in the future development of apparels (G.Cho,2010; Dunne,2007) there are still varieties of challenges in the research and practical applications (Dunne,2010) , which need to be further explored and promoted.

As a significant topic in the area of health care, the health surveillance of infants is also one of the important applications of smart clothing. Tai et al. (2008) attempted to put the sensors connected to the computer via Internet on the arm of infants so that the parents could monitor their child's health through a web browser. Coosemans et al. (2006) measured the electro-cardiogram signals of hospital infants by means of fabric sensors, demonstrating more comfortable than traditional sensors, but generally less accurate than conventional electronic devices. In short, smart clothing for infants shows a few of distinctive features different from other types of smart clothing as follows. Firstly, it needs to take into account more factors of safety and comfort since the infants are more susceptible to clothing than adults; secondly, it tends to be easily monitored and provided with timely feedback. As infants usually can't be able to give the feedback accurately on the basis of the measurement signals, the system is required to timely send the measurement results to the guardian. In such a process, the connection with the mobile device and Internet is usually necessary.

2.3.4 Discipline Domain of Smart Clothing

The emergence of smart clothing is caused by the common development of clothing design, material science, computer science and other disciplines, among which the multi-disciplinary cooperation is inseparable. The clothing designers' initiative of "wearable smart computers" has prompted material scientists to develop lightweight fabric sensors and integrated circuits, allowing computer researchers to design artificial intelligence programs on the basis of sensor signals. Then a number of analog signals of different types will be converted into the output mode that can be understandable and valuable, aiming to simulate human perception and processing of signals. The application of smart clothing in the specific fields requires the integration of more disciplines. For example, the smart clothing design for health surveillance can't be separated from the participation of medical experts. In short, with the continuous development of smart clothing, the requirements for multi-disciplinary cooperation are also constantly improving. However, this also increases the difficulty of cooperative research since it is inevitable to require many experts having an overall understanding of the relevant knowledge of different disciplines covering the in-depth investigation of clothing design, computer-based intelligent technology and intelligent fabric technology. As in the field of bioinformatics, the most successful researchers are often those experts in both of biology and computer science. Thus, this chapter will introduce the relevant knowledge in the main areas involved in smart clothing, linking these seemingly different fields.

2.3.4.1 Link with Clothing Design

Smart clothing is a new one in a blend of many disciplines and technology, hence it also requires integrating a variety of factors in the design process, such as electrical power, safety, comfort, art design and aesthetics. A set of scientific and universal design procedures are essential in a large-scale design and production process. However, there will be many new problems, for example, what is the order of design? How to design and ensure the least time consuming and cost? How to take into account the functional and aesthetic aspects at the same time? The emergence of these new problems brings new challenges in the field of clothing design. As the conventional clothing design process is no longer suitable for smart clothing, it's essential to conduct theoretical innovation and gradually sum up a set of design process suitable for smart clothing in a continuous practice.

Whether for a computer application system or industrial products, the first step is to pay attention to the demand analysis which refers to what users need and whether the computer can meet the demands. The literature divided people's demands for smart clothing into functionality, connectivity, persistence, maintainability, usability, producibility, wearability and affordability, and regarded smart clothing design as a clothing design system. Dunne et al. (2002) classified the demands as thermal management, humidity management, mobility, flexibility, size and applicability, durability, clothing maintenance and other aspects. The literature Aguiriano et al. (2003) defined smart clothing as digital apparels and divided the demands into persistence, ease of maintenance, comfort, safety, and artistic satisfaction.

In recent years, researchers have explored the general procedures of smart clothing design, and their studies mainly regard smart clothing design as a special design system of electronic products from the perspective of system theory. Ulrich (1995) compared the clothing design process with the traditional design process of electronic products, and attempted to replace several steps in the design process of electronic products into the steps of clothing design in accordance with the characteristics of smart clothing. The purpose was to achieve the unity of electronic products and clothing design. The study found that the main difference between clothing design and electronic product design was that there was a continuous feedback and iterative process in the design of electronic products or computer software. In other words, a prototype sample was firstly developed in a certain step, followed by testing and measurement in the subsequent steps, then constantly returned to the previous steps for improvement and perfection, and ultimately formed the final product. Only in this way could the modules achieve seamless connection with each other. However, this step was rare in the conventional clothing design, thus a number of technical details need continuous exploration and refinement in the future research.

Moreover, clothing designers need to take into account the design of electronic devices and wiring in the light of the characteristics of new apparels. They can not only ensure the normal operation of intelligent system, but also pays attention to the comfort, safety and other aspects of the clothing. This is a new challenge for clothing designers and it requires the teamwork of scientists from different fields so as to achieve the best results.

2.3.4.2 Link with Computer Science

As smart fabrics is not the equivalent of smart clothing, fully functional smart clothing must be able to conduct effective analysis of the data from the fabric sensors, and to "smartly" anticipate the changes that need to be made in the next step. This module belongs to the field of purely computer software, more precisely, belongs to the topics of artificial intelligence. The purpose of artificial intelligence research is to make computers, like humans, be able to respond to complex situations automatically and in a similar way to human beings. In the smart clothing system, the research level of artificial intelligence technology directly determines the level of "intelligence" of smart clothing. Furthermore, the computers need a variety of signals such as temperature, pressure, sound, humidity, the persons and location of signals, and other types of data, collected in line with the current environment. Then the most appropriate judgment will be made on the basis of these characteristics. For instance, a funny dress may be designed as different colors based on the moods of people. Suppose there are three moods: happy, sad and calm, which corresponds to the indicator lights of three different colors. Nevertheless, this is not an easy task for a computer since one's emotions are determined by very complex factors that are difficult to make sound judgments through simple rules. The best way to deal with this problem in the field of artificial intelligence is illustrated as follows. Firstly, it's necessary to collect varieties of relevant data such as pulses, brain waves, blood pressure, pressure data and sound of various parts of clothing, and other information, currently associated with the person by means of sensors. Secondly, it needs to convert analog signals into digital signals, and then carry out feature extraction using a variety of statistical machine learning methods. Thirdly, it's essential to establish classifiers in accordance with the historically empirical data, followed by determining the response which ought to be made on the basis of the results of classifiers. Usually the prediction accuracy of such problems can reach about 60% (Picard,2000). However, the difficulty will increase and the accuracy will be often reduced if the emotional category increases and the environmental factors become increasingly complex. At present, the technologies pertinent to machine learning, signal recognition, speech recognition in the artificial intelligence have made some progress in the past 20 years. Nonetheless, there is still a big gap compared with the level of human intelligence. In many issues, the predictive accuracy of smart computers is far below the accuracy level predicted by humans. Therefore, this problem will also become one of the bottlenecks in the development of smart clothing, which will largely limit the "intelligence" of smart clothing. In the future research, on the one hand, the researchers of smart clothing shall timely track the most advanced artificial intelligence technology, and apply the latest and best technology to the smart clothing; on the other hand, it's critical to develop the smart clothing applications suitable for the current level of artificial intelligence, which is to find out the appropriate applications even for the "lower IQ" clothing.

2.3.4.3 Link with Material Science

Smart fabrics have been investigated over 10 years, in which researchers constantly attempted to create the fabrics with a function of electronic components by virtue of various approaches. The electronic devices and embedded computers usually operate normally with the aid of electrodes and conductors. Moreover, the ideal smart fabric shall be a device that produces current (electrode) and transfers current (conductor), in the meanwhile, it shall also have a few of the features of clothing fabrics such as softness, comfort, safety, aesthetics and so on. Nevertheless, there is still a certain gap between current research and this goal, and it is difficult to find the basic material that can generate current, accurately sense the external environment and conduct current in the current fiber fabrics. Consequently, most of the research is the organic integration of electronic components such as sensors, electrodes and fabrics. It's significant to take advantage of the strengths of both the embedded computers and clothing as much as possible in a particular type of issue.

Many fabrics are electrically conductive (Tang,2006), such as conductive polymer, fiber, embroidery, or clothing products, etc. These fabrics play a significant role in the design of smart fabrics since they can be utilized as a way to transfer information or energy, so as to achieve the connection with the computers. The early smart fabrics were applied by Swallow and Thompson aiming to develop "fabric keyboards" based on "fiber fabrics of sensor"(Swallow,2006). The fabrics comprised three layers of fibers, of which the upper and lower were conductive, but the middle was not conductive. After the keyboard was pressed, the two conductive layers came into contact and then the switched-on circuit delivered the signals to the output of the keyboard, thereby realizing the function of the keyboard.

The more sophisticated smart fabrics are applied to detect the biological signals of the body (Catrysse, et al., 2004), such as heart rate, respiratory rate and so on. The sensors frequently used in hospitals are irritating to the skin and long-term measurements can impair the health of people. However, the smart fabrics can make up for this shortcoming. Van Langenhove et al.(2004) developed a "fabric electrode" for the measurement of electrocardiogram and heart rate. Such a material was to have the stainless steel fibers processed into knitted fabrics and then embedded in a tape wrapped around the chest, in order to attain the interaction with the human skin. The experiments showed that this sensor made entirely of fiber, could achieve the accuracy comparable to conventional sensors, although somewhat affected by the noise. This smart clothing could be used to monitor the patient's health, athlete's physical changes during exercise, and people's health in a heavy work environment.

Loriga et al. (2006) produced sensors and electrodes by virtue of electrically conductive and piezoresistive yarns, and had them woven into clothes for monitoring blood circulation. The blood pressure, pulses and other signals of the body measured by sensors would be converted to voltage output, and the electrode was made of stainless steel yarns and ordinary yarns. In the study of smart clothing, fiber optic technology has also been a widespread concern since it has the function of sensing and transmitting signals at the same time. Fiber grating sensors (Tian, et al.,2001) aim to measure the wavelength change in the surrounding environment due to stretching or temperature variations by means of the change in the refractive index of fiber kernel in a single mode, thus realizing the function of sensors. The fiber optic materials have great potentials and application prospects in the smart clothing design.

2.4

Summary

This literature review aims to fulfill two objectives.

Firstly, it investigates the Knowledge Science area. Knowledge science has a different way with other academic disciplines in strengths and features. The essence of knowledge science is how to recombine resource and knowledge-structure. It is also the essence of innovation. Knowledge science with a bird's-eye view on the advent of the knowledge-based society, and the spread of knowledge management. It can guide researchers to use the creative thinking methodology to make something new and design something to face the future. Our research aims to develop the interdisciplinary thinking approach of the fashion design education. Meanwhile, we apply the knowledge of knowledge science to support smart clothing design research work from a designer perspective, promote research and education in the field of Knowledge Science based on Fashion design and Smart clothing design.

Second, it investigates Fashion Design and the main design thinking methodologies in design education. The research has shown that the study of the design thinking approach in the fashion design field is still at the developing stage. The research of design and new technology has made uncoordinated contributions. The study lacks an integrated thinking approach to optimize efforts from different areas. The emergence of these new problems brings new challenges in the field of Fashion Design education. As the conventional Fashion Design education and clothing design process is no longer suitable for the social needs, it's essential to conduct theoretical innovation and gradually sum up a set of suitable for Fashion Design education and design practice. So, based on the existing situation we aim to establish a new strategical thinking approach for fashion design education. We show the details in the next chapter.

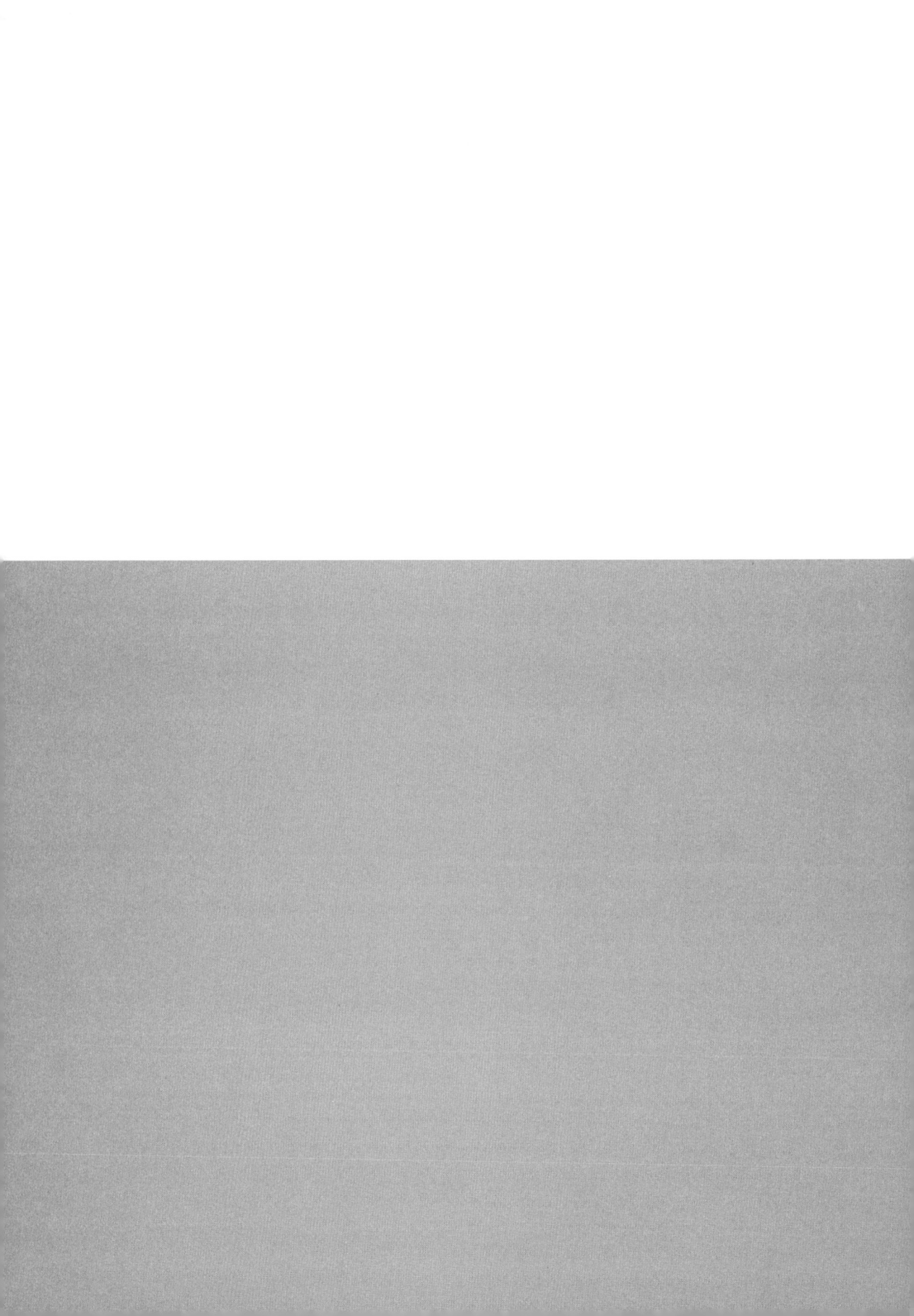

Chapter 3

Create New Thinking Approach
for the Fashion Design Education Field

3.1

Design Plus X: An Interdisciplinary Thinking Approach to Education Strategy in Fashion Design

Fashion design is a specialist field in which art and science are conspicuously combined, and the question of how closely and how organically art and technology are to be integrated is an issue. While technology emphasizes the point of demolishing the old to create new technology, art places importance on observing tradition to create new art. Thus, an organic fusion of art and technology includes the major issues of destruction and preservation, tradition and reform. Since fashion design education has the special characteristic of encompassing both art and technology education models, the major gap that has long existed between the sciences and the humanities must first be bridged in order to unite art and technology. For this purpose, it would be practical to assimilate multidisciplinary knowledge structures, to multilaterally reconsider from a space between tradition and reform, and to research a practical curriculum based on an education model that combines art and technology in fashion design with an emphasis on the formation and development of students' aesthetic sense, self-worth, and cultivation of the humanities. In developed nations, the concept of "art engineering" has already been proposed, and the model for cultivation of specialized human talent in fashion design has come to be highly evaluated as a model for cultivation of specialized human talent in art engineering. There is a preexisting fixed mode of the cultivation of graduate students majoring in fashion design in China, and the researches on the model for such cultivation are widely conducted. However, the results of such researches cannot be said to be sufficient. The lack of systematic research conducted on the fusion of art and multidiscipline from both theoretical and practical perspectives, namely, the lack of research relating to each stage of art engineering: its education model, system, methods, curriculum, teaching materials and practice, remains a major problem.

In regard to the current state of cultivation of specialized human talent in fashion design, the authors propose a new strategy for interdisciplinary practical education called DPX (Design Plus X), as an integrated new design philosophy based on multidiscipline crossover (Figure 3-1). "X" corresponds to knowledge from various crossover departments and fields (e.g. sociology, management science, electronic engineering, information science, etc.).

The purpose of the new practical system research called DPX is, with the whole process of design development as a carrier, to incorporate the knowledge from various crossover departments and fields represented by X into the design; and to utilize the broadening and deepening of knowledge related to design development- comprehensively multidisciplinary crossover fields such as Design Plus Humanities and

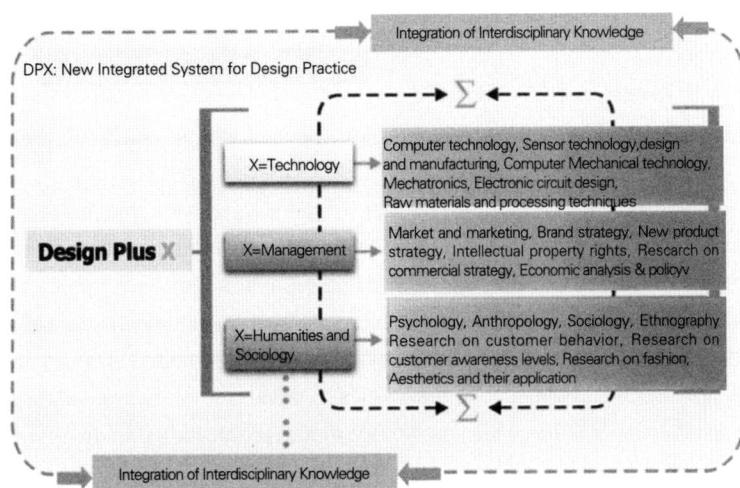

Figure 3-1 DPX: new integrated system for design practice

Sociology, Design Plus Management and Design Plus Technology; and to develop an integrated, innovative practical education model. Humanities-oriented research, the conception of commercial strategies and plans for technological realization, will be incorporated into practical education and practical training about integrated solutions related to product design and services. By doing so, the overemphasis on formal reorganization in art design education will be broken down, a core capacity for innovation that centers on reform of method and function will be cultivated, students' innovative practical abilities, abilities to integrate multidisciplinary knowledge, and the necessary attributes for entrepreneurship will be improved, and the cultivating effect in the practical stage of the art design will be enhanced. Concrete details are provided below.

3.1.1 Design Plus Humanities and Sociology

Design innovation is born from investigative research in the humanities and sociology . We target research on the impact of design on society and culture, as well as focusing on diversity in the marketplace, searching for customer needs, improving capabilities for product innovation, and evaluation of design results. Research on customer awareness levels and customer behavior, and knowledge from sociology and the humanities such as psychology will be incorporated into a new practical education for art design, students' innovative practical abilities and sense of social responsibility will be fostered, and students will build a new awareness of leading development in the industry.

3.1.2 Design Plus Management

The relationship between knowledge from management fields and specialized art design knowledge will be focused on, research related to profit acquisition models in the market will be merged with specialized art design knowledge, and knowledge of markets, marketing, brand strategy, intellectual property rights etc. will be incorporated into a new practical education for art design. Product development led by commercial strategy will be conducted, and market value will be added to products. The results of practical education will be pragmatically connected to the market, and market testing will be conducted. Students' entrepreneurial capabilities will be fostered.

3.1.3 Design Plus Technology

Focus will be placed on crossover between art design and various technological departments, and knowledge from technology fields such as mechanical design and manufacturing, electronic circuit design and computing will be merged with specialized art design knowledge. Engineering planning will be incorporated into product development, an engineering-oriented direction of research development will be presented, and realization of product function will be supported via engineering. Further reinforced practical capabilities will be cultivated so that students are able to connect design practice to the product manufacturing phase.

3.2

Design Practice Process of "DPX"

DPX is an important interdisciplinary practical strategy for fashion design education, and there is an increasing need to include Humanities and Sociology, management and technological considerations in strategy and learning planning processes in the future design. The design process of DPX has been designed to be flexible. It involves a series of four workshops and group presentation (Figure 3-2):

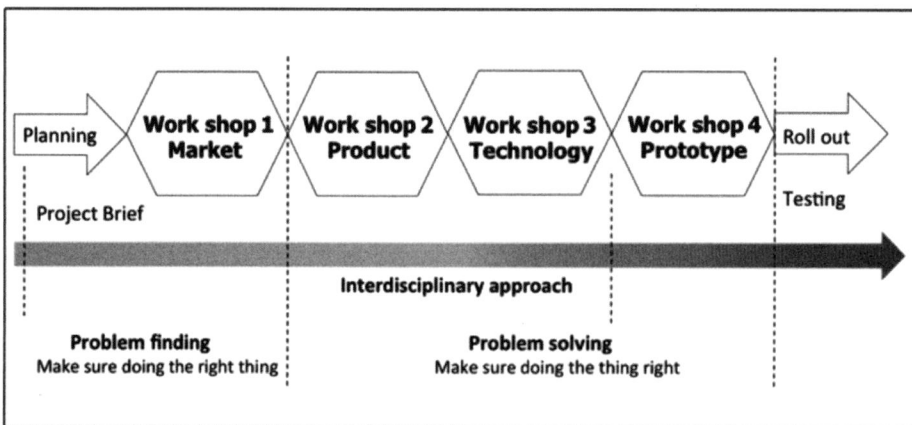

Figure 3-2 Design practice process of "DPX"

①Market: identification of market and business drivers.

②Product: generation of product feature concepts.

③Technology: identification of technology solution option.

④Prototype: refining or co-refining embodied solutions.

This is an iterative feedback loop that takes place until the desired outcome is achieved.

3.2.1 Workshop 1:Market

Dimensions of product performance are considered, market and business drivers identified, grouped and prioritised, for different market segments. The strategic context is considered and key knowledge gaps identified.

Understanding the user through observation and engagement that enables the design to articulate the problems that the user perceives. This draws on design ethnography, which we can describe as a way of understanding the particulars of daily life in such a way as to increase the success probability of a new product or service or, more appropriately, to reduce the probability of failure due to the failure to understand the basic behaviors and frameworks of artifact users.

Workshop 1

−Introduction the background of "DPX"

What is the knowledge Science?

What is the DPX thinking approach?

What is the Market of future design (fashion trend and fashion future)?

−Aim

Identify of market and business drivers.

Problems finding.

Design the project briefly.

−Method

Brainstorming / Questionnaires.

Workshop 1: 2days(8hours).

Team: each group of three students from Fashion School.

3.2.2 Workshop 2:Product

Based on workshop 1, analysis of the design problem, then choose one valuable design concept to solve the design problem. Based on the design concept, we need research on what kind of discipline knowledge we need to collect (Figure 3-1). Some best innovations come at the intersectional part of the different knowledge areas (Figure 3-3). Product feature concepts and key knowledge gaps are identified.

Workshop 2

−**Introduction**

   Introduction excellent design cases in different field.

−**Aim**

   Synthesize collected knowledge from different areas.

   Design innovation: get the new solutions/ concepts.

   Product features (sketch book and text description).

−**Method**

   DPX thinking approach.

   Workshop 2: 3days(12hours).

   Team: each group of three students from Fashion School; invite teachers from the different discipline.

   Research presentation: 20minutes.

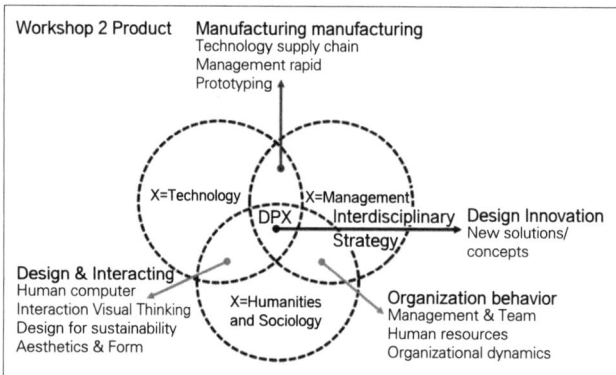

Figure 3-3 Design innovation by DPX thinking approach

3.2.3 Workshop 3:Technology

Based on workshop 2, generating or co-generating ideas around possible solutions. Technology options are identified, allow prioritization and ranking of product features and technology solutions (Figure 3-4).

Workshop 3

-Introduction

Smart tecnologies apply in diferent areas.

-Aim

Identify technology solution.

Define the design/technical terms.

Clarify the contributions of both areas.

Experiment: data and methods.

Design illustration.

-Method

DPX thinking approach.

Smart Technology.

Workshop 3: 5days(20hours).

Team: each group added two new students.

Three students from Fashion School and two new students from information school.

Invite teachers from the information school.

Research presentation: 20minutes.

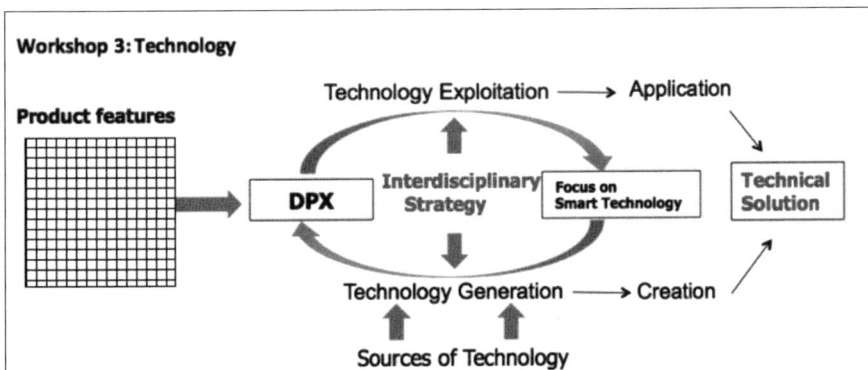

Figure 3-4 Technical solution by DPX thinking approach

3.2.4 Workshop 4:Prototype

Based on workshop 3, the team co-refining embodied solutions. Producing rapid prototypes or mock-ups of the selected sub-set of possible solutions. This is an iterative feedback loop that takes place until the desired outcome is achieved (Figure 3-5).

Workshop 4

−**Aim**

Make the prototype.

Testing.

Summarise outputs from the workshop.

−**Method**

DPX thinking approach.

The method of clothing design.

Ten principles for good design.

Workshop 4: 10days(40hours).

Team: each group added two new students.

Three students from Fashion School and two new students from Information School.

Research presentation: 20minutes.

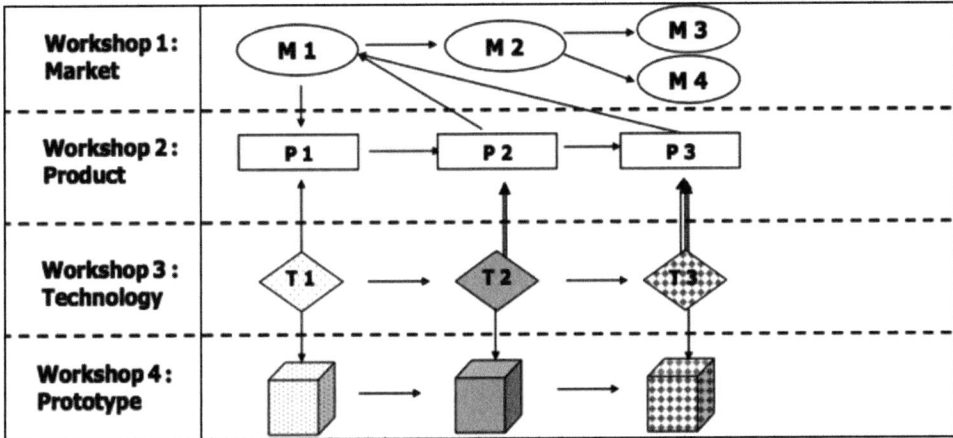

Figure 3-5 Design iterative feedback loop by DPX thinking approach

3.3

Keeping the Process of "DPX" Alive

The "Design Plus X" is new strategical thinking approach for fashion design education. It could offer the directions for the creative boundary extension and enhance student capability to realize what is co-design by himself in his own context. It could clarify the roles of all participants (designers and engineers) within the collaborative development teams. "DPX" brings a fresh set of attitudes, aspirations, and capacities. It provides the expertise, sets new standards that others will rise to, and contributes to the development of capable and creative designers.

We try to keep "DPX"alive in the practice of Fashion Design, "X" corresponds to knowledge from various crossover departments and fields. "X" is changeable and flexible, it makes the designers be more innovative (namely, Inspired Designer) to find the new area of Fashion design. Here we use smart technology as the case (Figure 3-6).

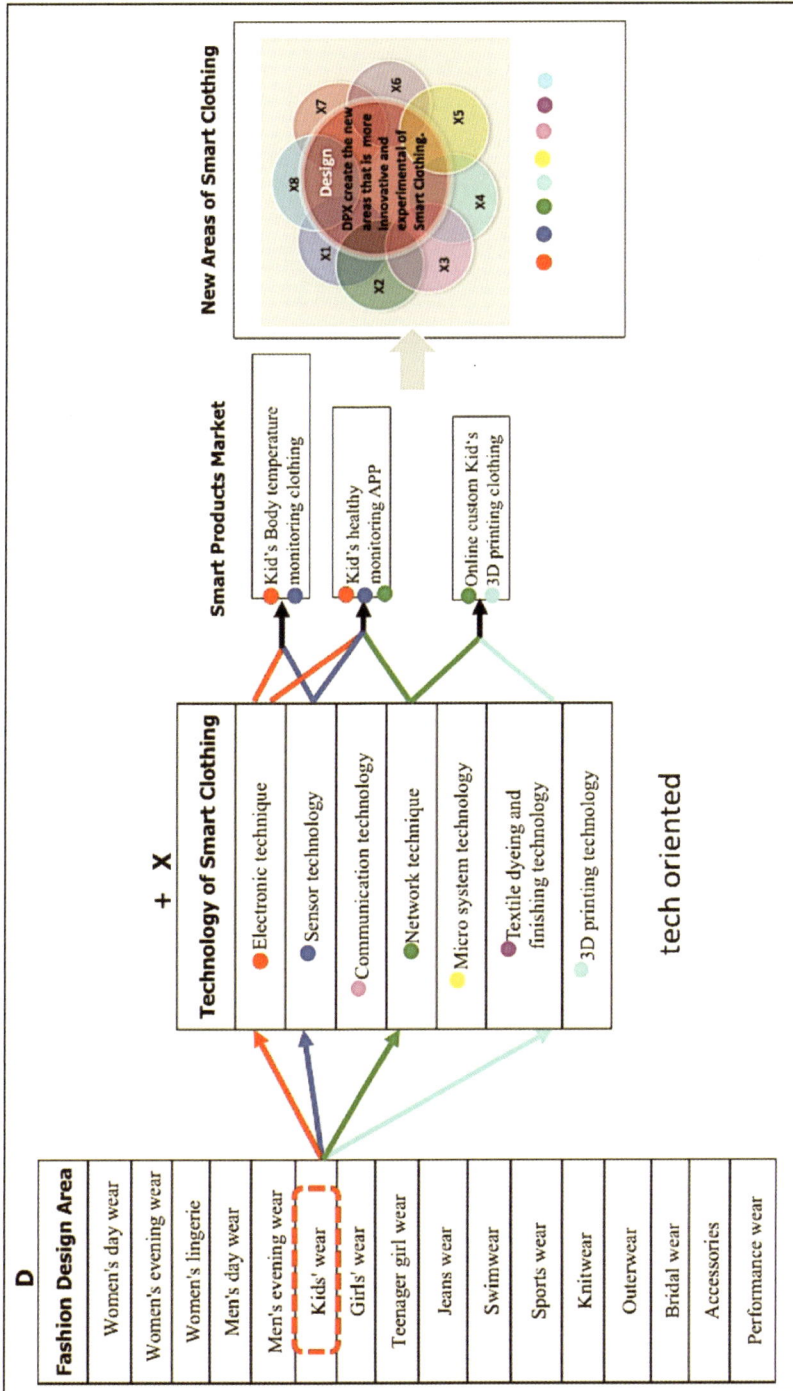

Figure 3-6 DPX: strategies for the new areas of fashion design field(take smart technology as cases)

3.4

Summary

(1)Resolving overemphasis on formal reorganization in design education.

With the whole process of design research development as a carrier, multidisciplinary fields will crossover comprehensively such as Design Plus Humanities and Sociology. Design Plus Management and Design Plus Technology, to develop a new practical method, and by enforcing the consistency of DPX, a new integrated concept for design practice within practical education, overemphasis on formal reorganization in design education can be resolved, and a core capacity for innovation that centers on reform of method and function will be cultivated.

(2)Improvement of students' innovative practical abilities, abilities to integrate multidisciplinary knowledge, and the necessary attributes for entrepreneurship.

Design Plus X is a new integrated concept for design practice based on multidisciplinary crossover. Focusing on the reform of design and the reform of design application, through humanities and sociological research, searching for customer needs, and technology and service planning, a new practical method for art design will be developed via multidepartment and multidiscipline crossover, and students' innovative practical abilities, abilities to integrate multidisciplinary knowledge, and the necessary attributes for entrepreneurship can be improved.

(3)Formulation of an educational strategy via a new practical education curriculum.

An educational strategy will be formulated via design practice modules that are appropriate for the development of education in art design departments (regular course) of normal higher education institutions, including a practical education and cultivation model, practical education curriculum, and practical education subject contents.

(4)Solve the problem of fashion education knowledge structure.

Through educational reform via Design Plus X, problematic points such as overemphasis on formal reorganization in art design education, the value of core innovation that centers on reform of method and function in design development, neglect of the cultivation of a cultural spirit, and the lack of sociological knowledge such as research on customer awareness levels and customer behavior, will be thoroughly resolved.

(5)Solve fashion practical teaching and learning problems.

Through reform of practical education subjects, the overemphasis on cultivation of manual work abilities in art design educational practice will be concretely resolved. Taking product design as an example, conventional design practice is merely a process leading to the pattern production phase, and it has not been possible to connect design practice to the manufacturing phase as it has not incorporated multidepartmental knowledge; as a result, it has not been possible to conduct extensive testing of design feasibility.

(6)Establish a new fashion practical teaching management model.

Through the DPX New Team Formation Plan, via the implementation of a practical education curriculum administration model based on cooperative competition and mutual growth, the issue of the absence of an effective method for implementing innovative practical education will be resolved, and in addition to presenting an effective method for implementation of practical education, the cultivation effect during practical stages will be enhanced, and pragmatic effects due to the integration of practical stages and the market can be improved, as well as students' motivation to learn.

(7)Establish a DPX practical learning website for students.

By making students participate in website construction, student autonomy will be proactively elicited, independent learning abilities will be cultivated, and cohesiveness as a team will be formed. As a result, the issue of the absence of a method for eliciting student assertiveness in practical education will be resolved.

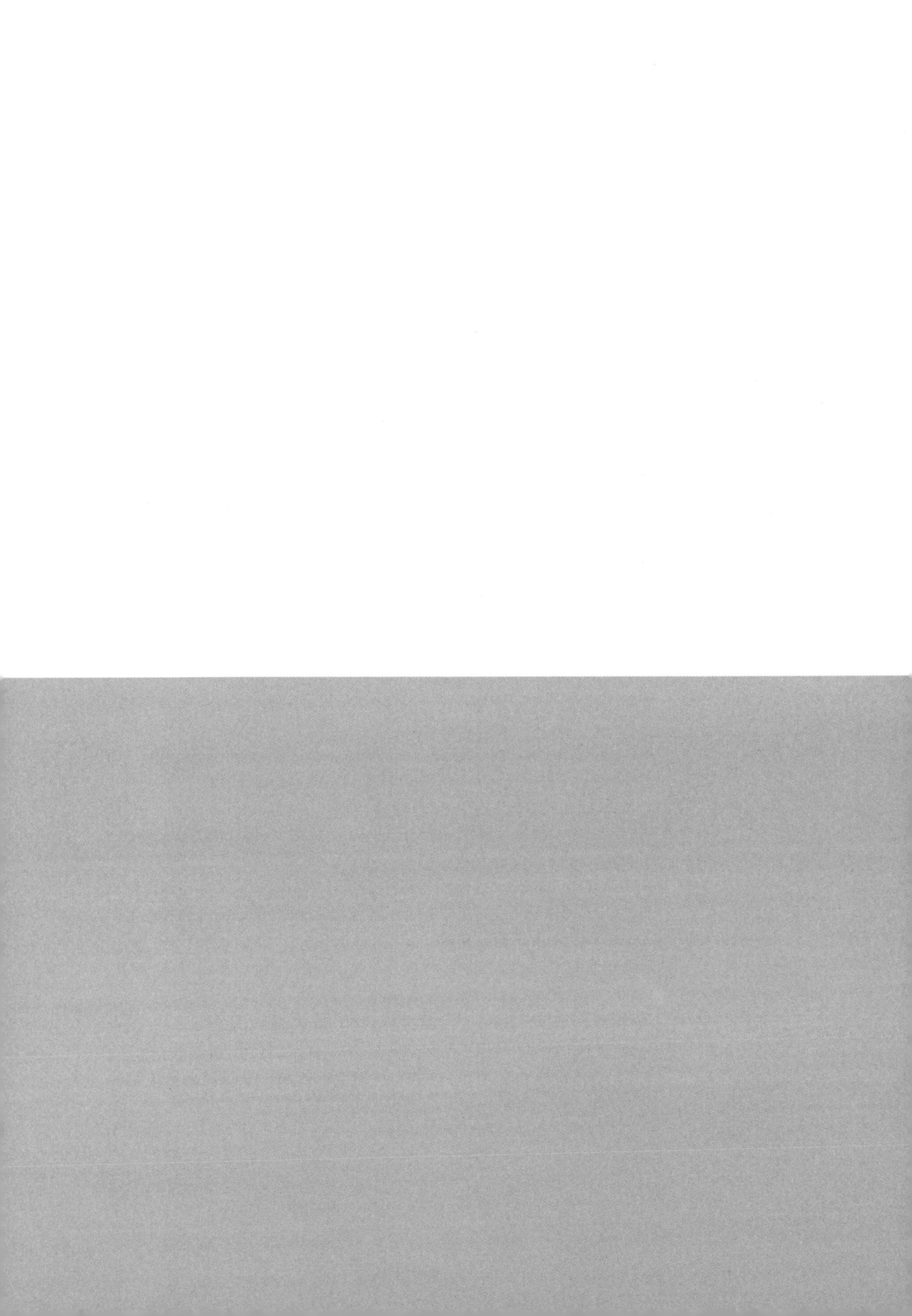

Chapter 4

The DPX Case Studies in
Fashion Design Education:
New Strategy for Interdisciplinary Practical Education
with Specific Focus on Smart Clothing

4.1

Case Study 1

Project: Designing Comfortable Smart Clothing: for Infants' Health.

Thinking Approach: Design Plus "X" .

Team members: DingWei (Fashion Design), Yukari Nagai (Knowledge Science), Liujing (Fashion Design), GouXiao (Information Science), XiaoLin (Information Science).

"DPX" workshop process: Figure 4-1.

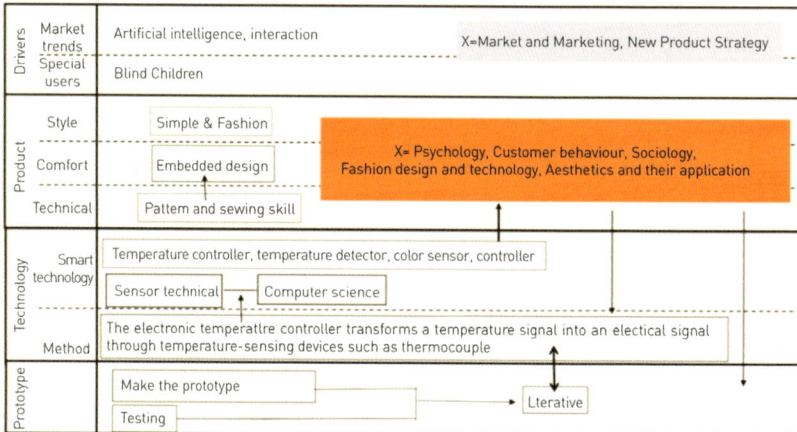

Figure 4-1 "DPX" workshop process

4.1.1 Research Introduction

Well-being is a major goal of human society and its realization is a challenge for designers. Recently, health-conscious and new technologies are being developed to support the healthy lifestyle of individuals. Clothing is an essential item that is required to wear daily. Clothing is unique because it is personal, comfortable, worn directly on the body and used globally (Kirstein, et al., 2005). Clothing is an ideal location for intelligent systems as it enhances "our capabilities without requiring any conscious thought or effort" (Mann, 1996). Smart clothes utilize sensors and artificial intelligence programs to respond smartly to signals emanating from the human body and the environment. They can be used, for example, to monitor the health condition of people in hospitals or at home by measuring vital index, such as body temperature, blood pressure and heart rate (Rai,et al., 2012; Lymberis & Olsson, 2003; Malins, et al., 2012). They can also be used to measure the physical condition of people working under extreme conditions (Rantanen, et al., 2000; Van Langenhove & Hertleer, 2004).

Health monitoring is an essential application of wearable sensor systems, especially for infants. Wearable sensor systems have enabled the creation of a new generation method of constant health monitoring for infants (Zhihua, 2015). Traditionally, infant health is monitored under the direct supervision of clinicians and parents. This method requires dedicated manpower, and sometimes, it is difficult for clinicians and parents to identify the infants' potential physiological condition. The initial purpose of the application of wearable sensor systems in infant health monitoring is to provide them everyday-centered care to ensure that appropriate care or therapy can be given at the onset of complications, thereby reducing the cost of hospital-based clinical interventions and the burden on parents at home. A health monitoring system is therefore not only a tool that provides an early indication of changing patient status allowing for early intervention but also a means by which the effect of interventions and therapies may be recorded, evaluated, and controlled (Murković, 2003).

One of the goals of smart clothing design research is to develop sensor technologies that are more comfortable to wear and hygienic (washable), thereby facilitating their everyday use and perception as second skin (Axisa, et al., 2005). In smart clothing, human aspects are derived from the integrated characteristics of clothing and electronic devices. As a result, smart clothing can simultaneously provide the usability and functionality of electronic devices as well as the comfort and fashion of clothing. Additionally, smart clothing can provide safety and durability; traits that are common to clothing and electronics (Gilsoo Cho, 2010). Temperature is a key vital sign used by clinicians, parents and caregivers in order to assess infants during times of acute illness. To achieve long-term and accurate monitoring of infants' core temperature, most temperature sensors are integrated into infants' clothing. The figure1 shows the framework of our approach for the design of a health monitoring system that considers well-being and incorporates a wearable sensor and smart technology. It can be applied by adapting a temperature sensor and machine learning. Based on this framework, we propose the design of a health monitoring system for infants with a focus on comfort. Previous studies on smart clothing had limitations in terms of comfort because of the number of sensors needed to capture body temperature (Dunne & Smyth, 2007). Understandably, comfort is one of the biggest challenges in smart clothing design (Dunne, 2010; Dunne, et al., 2007) conducted an experiment to compare the accuracy of different styles of smart clothing in detecting bending movement. They found that sensors on tight-fitting clothing had increased measurement accuracy but were less comfortable than sensors on loose-fitting clothes. They noted that the trade-off between sensor accuracy and wearable comfort was one of the main challenges in the design and commercialization of smart clothing.

4.1.2 Research Questions

As mentioned above, the comfort of smart clothing is one of the largest challenges in the development of smart clothing, and the comfort of clothing usually shows contradictions with the measurement accuracy. In a general design process, the comfort will tend to be overlooked if the focus is on the accuracy, and vice versa. Scientists also attempt to improve the comfort of smart clothing by virtue of various methods, such as the investigation on intelligent fabrics. However, the current research still has a lot of limitations, hence this thesis presents a new solution in response to this problem. Such a solution, different from smart fabrics that focus on hardware changes, aims to improve the accuracy as much as possible with the aid of the design of artificial intelligence algorithms on the basis of maintaining the comfort of clothing, so as to achieve the organic unity of the comfort and accuracy. Then this thesis will introduce the research

Table 4-1 The measures for comfort quality evaluation

Dimeusion Item	Degree	Description
Mood	High / Low	Worried about wearing unpleasant and nervous
Appendage	High / Low	Feeling the electronic devices on the body
Injury	High / Low	Causing injury or pain to the human body
Sense of change	High / Low	Feeling some change and strange
Activity	High / Low	Physical activity affected by devices
Anxiety	High / Low	Feeling hidden danger

pertinent to the comfort of clothing, illustrate the contradiction between the comfort and accuracy of sensors in details, and the solution to such a problem.

4.1.2.1 The Comfort of Clothing

The comfort is a significant indicator of clothing design since wearing comfortable clothing is one of the main functions. People can easily understand that the comfort is a very intuitive indicator. Nevertheless, scientists attempt to quantify the comfort index in the study of clothing design, and thus carry out sub-research by virtue of mathematical tools and conduct a complete modeling analysis combined with other indicators in the clothing design.

4.1.2.2 Quantitative Research on the Comfort

Barfield et al.(2001), conducted systematic investigation on the theories pertinent to the comfort and safety in smart clothing design, and put forward some theories and principles. As smart clothing was closely linked to electronic components, and served as the intelligent system between clothing and computers, the measure of the comfort of clothing appeared much different from that of general clothing. Knight et al.(2005), proposed a series of specific quantitative principles for smart clothing and wearable computers. They divided the indicators measuring the comfort of smart clothing into six dimensions consisting of mood, appendage, injury, sense of change, activity and anxiety, as shown in Table 4-1. The degree of each indicator could be obtained through questionnaire, and then these indicators were unified and used for a complete modeling analysis combined with the conventional evaluating indicators of intelligent systems. These indicators aimed to study smart clothing by means of mathematical methods. Although these indicators are still not perfect at present and have not been widely applied in industry in the next few years, they are still meaningful innovation. Actually smart clothing not only belongs to the computer system and shows closely linked with the computer performance such as accuracy, but also belongs to the apparel industry and appears inseparable from the daily life of people and commercialization. Therefore, it's critical and essential to quantify the comfort, safety and other non-conventional evaluation indicators, and then conduct a complete modeling analysis combined with the conventional evaluating indicators of computers such as accuracy. In such a way, it will be more conducive for scientists and businesses to develop and produce accurate and comfortable clothing.

4.1.2.3 The Comfort in Infant Clothing Design

As infant clothing is very different from adult clothing, the infants have immature skin and appear lack of self-protection awareness, and their physiological function is not fully developed. Coupled with other factors, hence infant clothing has much more stringent requirements than adults in safety and comfort. For instance, it's common to utilize more soft cotton fabrics in the infant clothing design, of which the pattern design requires minimizing segmentation. What's more, the treatment on the decoration, buttons and other hard parts ought to be particularly careful for infants so as to avoid accidental scratches or ingestion. In comparison, smart clothing also contains a number of devices with such potential security risks, which brings great challenges to smart clothing design. Besides, the sensors closely attached to the skin of infants can only be used for a short period in the medical measurement, but can't be used in daily life as real clothing. However, due to their lack of expression and feedback on physical discomfort, the infants are one of the most marketable users of smart clothing. The accurate and comfortable smart clothing for health surveillance will greatly improve the health of infants, in the meanwhile, it will reduce the difficulty of guardianship of parents. As a result, it will become the significant trends of the apparel industry in future. However, there are rather few related studies at present because of many factors. Firstly, the comfort of smart clothing for adults has become a large challenge, thus smart clothing for infants will be more demanding. Secondly, as infants are far weaker than adults in all respects, the research and experimentation will be more risky, greatly hindering the development of smart clothing for infants.

4.1.2.4 Definition of Research Questions: the Conflict Between the Comfort and Accuracy

One of the main objectives of the investigation on smart clothing is to make electronic components into the apparels that can be worn with comfort and wash-ready in daily life, and even become "the second layer of the human skin". As it is different from electronic components and computer facilities, the smart clothing design requires more attention to the characteristics of clothing, such as comfort, re-usability, safety, aesthetics, fashion, commercialization and so on. It's not just bundling electronic devices to the body or serviced to the specific persons in special occasions. Moreover, the comfort is not only one of the most significant indicators, but also one of the largest challenges in smart clothing design (Dunne, 2010). For example, designing a shirt that can monitor body temperature in real-time usually requires keeping the temperature sensors close to the sensitive parts of the skin, such as armpits and mouth. Nonetheless, this will make people feel uncomfortable, hence it only applies for the timing measurement in hospital, but not applicable for the real-time monitoring in daily life (Table 4-2).

Dunne et al.(2007), designed experiments to compare different styles of smart clothing for measuring body movements, and found the clothing closely attached to the skin usually required high measurement accuracy while the measurement accuracy of relatively loose clothing appeared significantly reduced. The main reason was that the human motion was measured by means of pressure sensors, however, the loose clothing made the pressure sensors unavailable to closely contact the body, resulting in the noise of the pressure sensors' signal. Taking this experiment as an example, Dunne et al. summarized one of the major challenges in the current smart clothing research, which was the contradiction between the comfort of clothing and measurement accuracy of sensors.

Table 4-2 Issues: difficult to address both comfort and sensor accuracy in real-word application

Temperature Sensor	Weight	Size	Signal	Feel
Skin–tight sensors	Heavy	Big	Accuracy	Uncomfortable
Non–contract Sensor	Light	small	Inaccuracy	Comfortable

4.1.3 The Traditional Solution

In fact, scientists have been studying the issue for many years and have tried different approaches. One of the hottest topics is smart fabrics, which, as mentioned above, aim to make the conductive fibers into the fabric-based sensors and wires, so as to achieve the same comfort as regular fabrics (Cho,2010; Coosemans,2006; Jang,2007). Nevertheless, there is a problem that these sensors on the basis of conductive fabrics are still unable to achieve the same accuracy as conventional metal sensors, and they still need to cling to the skin in many cases. Smart fabrics change just to make the sensors soft, but still do not alter the requirements for smart clothing to have sensors closely contact the skin.

Furthermore, the fabric-based sensors are more expensive than traditional metal sensors, largely limiting the commercialization process. Another simple way to improve the comfort is to utilize the non-skin-tight sensor, which refers to some space between the sensor and the skin, as shown in Figure 4-2. Intuitively, the signal measured by this sensor loses accuracy. For instance, if the thermometer is placed 5mm from the skin, or another layer of clothes separates the thermometer from the skin, the displayed temperature will be rather different from the skin thermometer. In the field of electronics and computers, such non-contact sensors are usually less investigated for reasons of accuracy loss. However, in the field of smart clothing, these sensors can play its role.

Dunne et al. (2007), tested the effectiveness of different types of non-contact sensors in the smart clothing for identifying human postures in experiments, and found that the more relaxed and comfortable the experiments were designed, the lower the measurement accuracy of sensors showed. Although the non-contact sensors were applied in this study, the accuracy of the non-contact sensors was not improved using this method. Consequently, it was almost unavailable to solve the conflict between the comfort and accuracy. The next question is how to improve the accuracy of the non-contact sensors? One of the most expensive solutions is to install infrared devices on the clothing, but its defects are also obvious. At present, infrared devices can't reach the level of real-time measurement, and will lead to potential security risks. Another way is to create a function that describes the relationship between the temperature of the non-contact sensors and real body temperature. Nonetheless, it needs to obtain a number of environmental parameters, such as the distance between the sensors and the skin, the postures of the body, the motion parameters and other parameters. As these parameters vary with the movement of the human body, it usually makes it difficult to acquire accurate data, thus appearing challenging to describe the function accurately. Therefore, the traditional solution has great limitations.

4.1.4 The Solution of the Project

The objective of this paper is to contribute to the literature on smart clothing design by proposing a creative solution that can help address the trade-off between accuracy and comfort in smart casual clothing (Ding, et al., 2015). We used the "Design Plus X" thinking model ("X" means different areas of knowledge or the technique) to synthesize the collected knowledge and solve research issues. For this project, "X" refers to both sensor technic and computer science. We employed the "DPX" thinking approach to design an experiment for body temperature monitoring using smart clothing and with enhanced sensor performance. Knowledge of clothing design, temperature sensor technology, and machine learning was applied in the experiment to test the hypothesis that multiple skin-loose sensors on casual clothing can effectively be used to estimate the body temperature of the wearer. The study thus presents a creative solution that can be an important stepping-stone toward the design of comfortable smart clothing for health monitoring, driven by the socially-oriented motive of design for well-being (Nagai, 2014), as Figure 4-2 and Figure 4-3 shown.

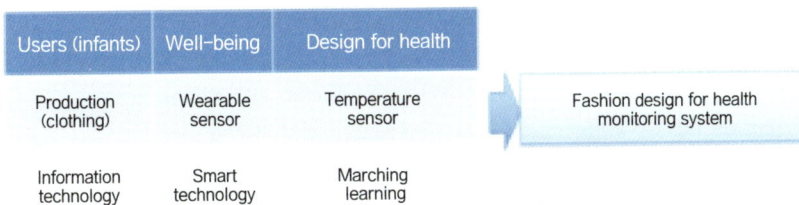

Figure 4-2 Framework of fashion design for health monitoring system

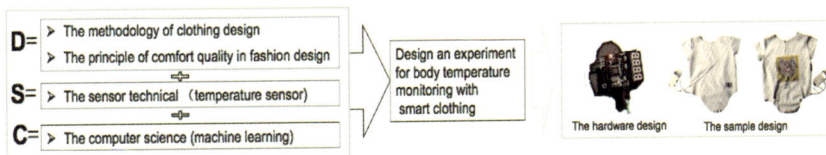

Figure 4-3 The "Design Plus X" thinking approach in smart clothing design

4.1.5 Experimental Design

4.1.5.1 Measuring Body Temperature

It is well known that the measurement of human body temperature plays a very important rde in health monitoring, either in the hospital or at home. The change in body temperature can be used as an indicator of various illnesses. Classical methods to measure body temperature can be categorized into the following types, according to the body parts measured (Robinson, 1997; Kelly, 2006):

(1)Measurement inside the body (e.g., oral, rectal, and gut using temperature sensors), which can get very close to the core body temperature but are obviously difficult to apply to smart clothing because clothing is not usually worn on these parts of the body.

(2)Using sensors that come into direct contact with the skin (e.g., under the arm), which can obtain an approximation of body temperature. This is commonly used for health monitoring at home as well as for smart clothing because of its simplicity. The disadvantages, however, are the inaccuracy and discomfort when in close contact with the human body.

(3)Using an infrared thermometer, which detects body temperature based on a portion of the thermal radiation emitted by the human body. While, this is a non-contact thermometer with high accuracy, there is a potential safety problem attributed to the radiation from the device. It is also too complex and heavy to be a component of clothing (Kay Wang, Peter Gill, et al., 2014). Overall, the major shortcoming of this method is the reduced quality in comfort when embedded as part of the garment.

All in all, the major drawback of these methods is the discomfort that they cause when used as part of a garment. We considered electronic temperature sensors that did not come into contact with the skin, i.e., the sensors located either in the space between the skin and clothing, or between two layers of cloth (Figure 4-4). Intuitively, the temperature measured by these sensors is correlated to body temperature, but tends to be affected by noise. For example, part of the information would be lost during the transmission in the space between the cloth and skin. Nevertheless, the signals obtained by these sensors can provide better information than a random guess, and can be viewed as a weak predictor. Our basic idea is to investigate whether the combination of several weak predictors can produce a strong predictor; that is, can the temperature measurement performance be further improved by combining different signal types from different sensors.

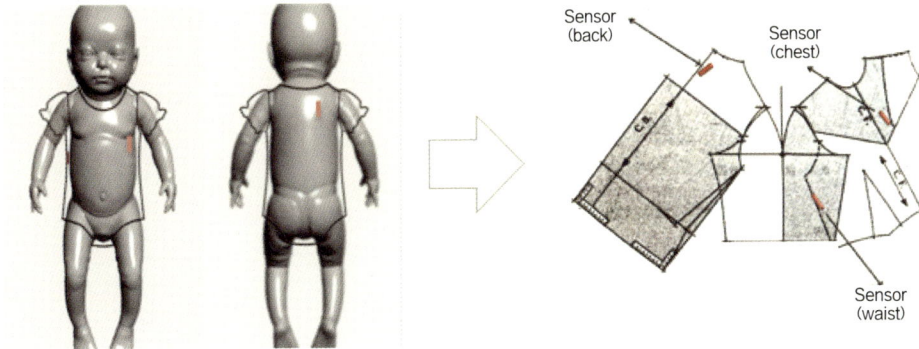

Figure 4-4 Garment pattern with sensors for smart clothing design

4.1.5.2 Machine Learning Framework

A smart clothing was designed in this study with two functions: the prediction of body temperature and warning of high body temperature. The same hardware was used for both purposes. The clothing was equipped with three temperature sensors positioned at locations corresponding to the chest, back, and waist. The sensors were placed on the inner surface of the clothing but not attached to the skin. To imitate daily activities, the subjects who wore the shirt assumed various positions to include sitting, standing, walking, and lying down, with the sensors alternating between touching and not touching the skin. The outputs of the sensors were thus rather unstable and noisy. The temperature measurements were recorded by each sensor every 2 seconds. Another temperature sensor was also used to monitor the room temperature because environment can significantly impact the performance of non-contact sensors. Different types of signals corresponding to heterogeneous conditions were obtained for the purpose of determining the best individual sensor and whether it is possible to improve accuracy by combining the three sensors. A series of standard step-by-step processes of supervised machine learning were used (Bishop & Nasrabadi, 2006), including data annotation, feature extraction, training and testing (Figure 4-5).

We designed various experiments to examine the impact of different sensors (Table 4-2), feature extraction methods (Table 4-3 ~ Table 4-5), regression models (Table 4-6), and sizes of training data (Figure 4-6).

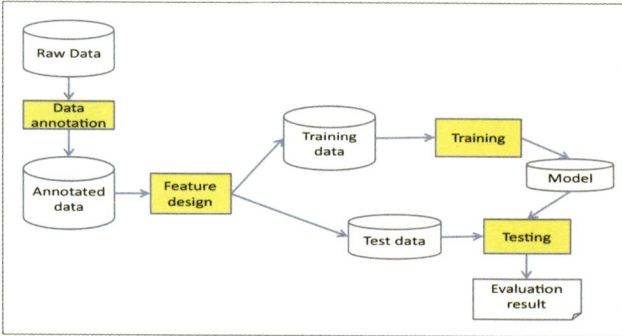

Figure 4-5 The machine learning process within the smart clothing system

Table 4-3 Performance of sensors from different body parts

ID	Sensor–chest	Sensor–waist	Sensor–back	R^2
1	No	No	No	0*
2	Yes	No	No	34.93%
3	No	Yes	No	5.99%
4	No	No	Yes	14.49%
5	Yes	Yes	No	33.52%
6	No	Yes	Yes	12.34%
7	Yes	No	Yes	26.49%
8	Yes	Yes	Yes	29.80%

Notes: *Simple average method.

Table 4-4 Impact of features from different time intervals

Time Interval (min)	R^2
1	30.86%
2	32.66%
3	20.80%
1+2	**34.93%**
1+2+3	33.23%

Table 4-5 Performance of different types of features

Features	R^2
Average temperature	13.50%
Average temperature + Variance	20.57%
Average temperature + Environment temperature	27.93%
Average temperature + Variance + Environment temperature	**34.93%**

Table 4-6 Performance of various kernels

Kernel	R^2
Linear (c = 0.01)	4.22%
Polynomials (d = 3, c = 0.1)	29.85%
RBF (g = 0.1, c = 0.5)	**34.93%**
Sigmoid (c = 0.5)	31.72%

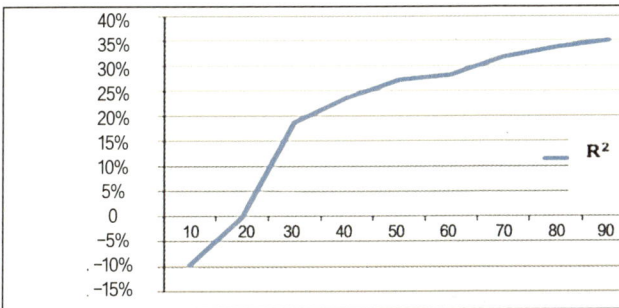

Figure 4-6 Learning curve with different amounts of training data

The evaluation measure used is the classical metric for regression evaluation the coefficient of determination or R^2, which measures how well a regression line fits a set of data, with the value of 1 for a perfect prediction and 0 for the simple average method. For 100 examples, we used a 10-fold cross-validation technique for evaluation; that is, the dataset was divided into ten groups and, in each of the 10 rounds, one group was used for testing and the others were used for training. In the experiment, to obtain stable results given the small sample size, we repeated the cross-validation 10 times and used the average performance as the final evaluation score.

Data annotation: because machine learning involves the learning of historical data and using the knowledge to predict future data, a number of manually annotated data are usually required for training and evaluation. In this study, we labeled 100 examples over 10 days. Each example was related to a specific point in time at which the body temperature was measured by a thermometer. These points include body temperature, measured using a thermometer under the arm and considered as the gold standard. These points also include a feature vector derived from the signals of three thermal resistive sensors placed at the chest, waist and back. The temperatures measured by the three temperature sensors as well as the environment temperatures were recorded. During the experiment, one individual, a three-year-old girl, wore clothing with the three sensors. During sampling, the body temperature and sensor signals at particular moments were recorded. The data was sampled across ten days, and there were about ten examples per day as depicted in Figure 4-7~Figure 4-9. We used degree centigrade as the unit for temperature measurement in the

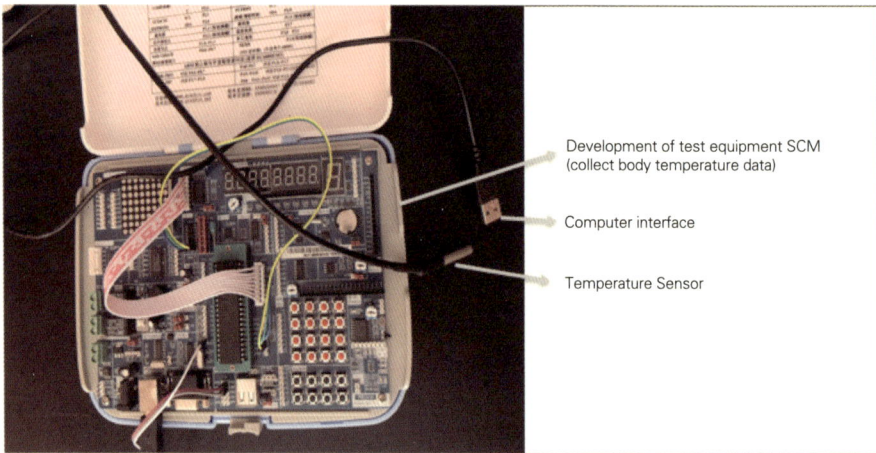

Development of test equipment SCM (collect body temperature data)

Computer interface

Temperature Sensor

Figure 4-7 The experimental photos

Smart Clothing Data annotation

No.	Sensor-chest(°C)	Sensor-waist(°C)	Sensor-back(°C)	Real Boday (°C)	Environment (°C)	Date
1	33.63	32.44	31.5	36.2	21	2016.05.30
2	33.5	32.56	31.56	36.2	21	2016.05.30
3	33.56	32.63	31.63	36.2	21	2016.05.30
4	33.75	32.69	31.75	36.2	21	2016.05.30
5	32.81	32.56	32.63	36.2	21	2016.05.30
6	32.44	31.5	32.5	36.2	21	2016.05.30
7	32.56	31.56	32.56	36.2	21	2016.05.30
8	35.63	31.63	32.75	36.2	21	2016.05.30
9	32.69	31.63	32.81	36.2	21	2016.05.30
10	32.56	31.75	32.44	36.2	21	2016.05.30
11	31.5	32.5	32.56	36.2	21	2016.05.30
12	31.56	32.56	32.63	36.2	21	2016.05.30
13	31.63	31.75	31.56	36.2	21	2016.05.30
14	31.56	32.63	31.63	36.2	21	2016.05.30
15	31.63	32.5	31.75	36.2	21	2016.05.30
16	31.63	32.56	32.63	36.2	21	2016.05.30
17	31.75	32.75	32.5	36.2	21	2016.05.30
18	32.5	32.44	32.81	36.2	21	2016.05.30
19	32.56	32.56	32.44	36.2	21	2016.05.30
20	31.75	35.63	32.56	36.2	21	2016.05.30
21	32.63	32.69	35.63	36.2	21	2016.05.30
22	32.5	32.56	32.69	36.2	21	2016.05.30

Figure 4-8 Smart clothing manually annotated data annotation

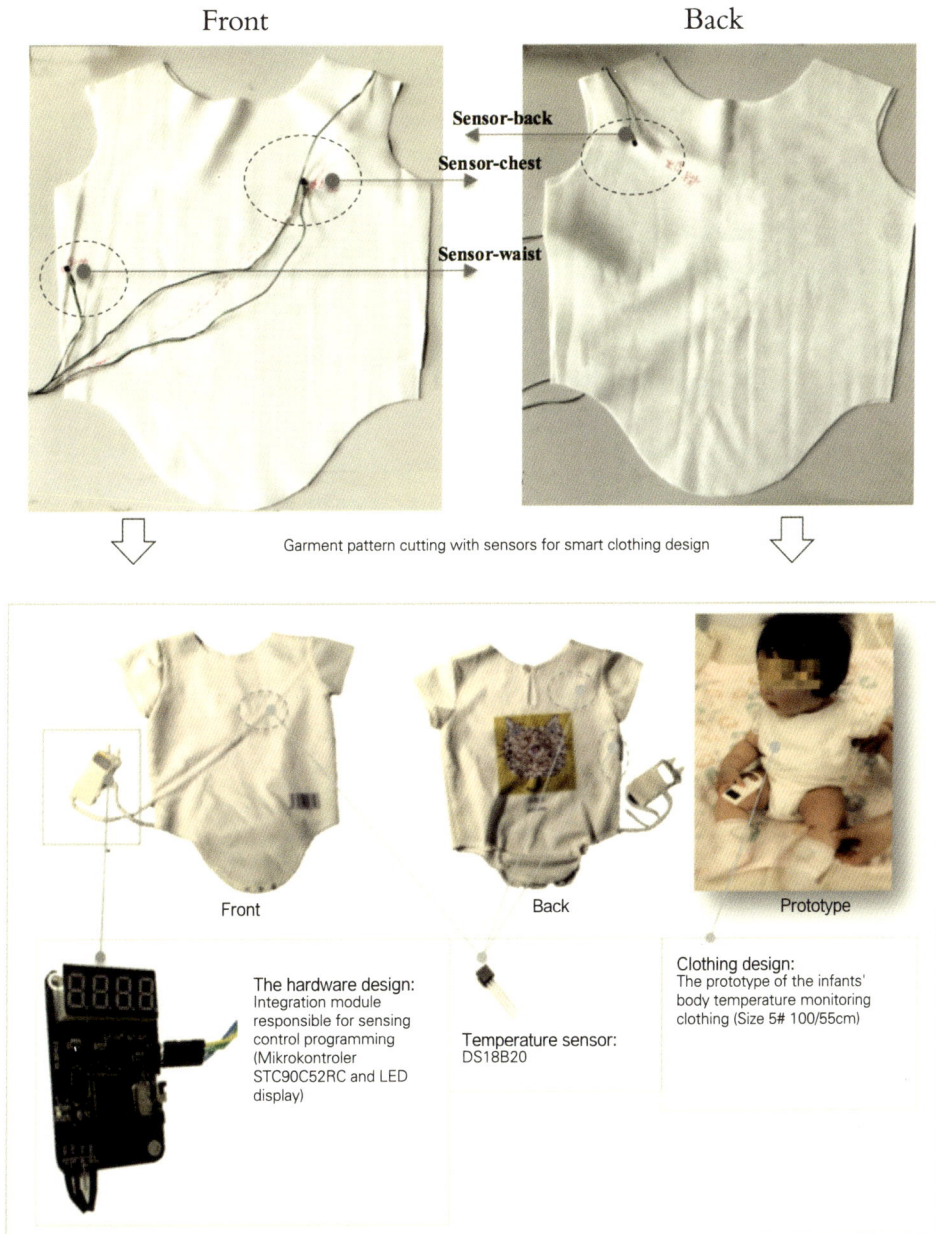

Front   Back

Sensor-back

Sensor-chest

Sensor-waist

Garment pattern cutting with sensors for smart clothing design

Front   Back   Prototype

The hardware design:
Integration module
responsible for sensing
control programming
(Mikrokontroler
STC90C52RC and LED
display)

Temperature sensor:
DS18B20

Clothing design:
The prototype of the infants'
body temperature monitoring
clothing (Size 5# 100/55cm)

Figure 4-9 The experimental photos

experiment. To increase the diversity of the examples, the time interval between two samples was set to at least 5 minutes, and the temperatures of the sensors were recorded every 2seconds before or after the sampling point. Hence, in the temperature regression task, the gold standard was the body temperature measured by the thermometer, and the features were extracted from the signals of the three skin-loose sensors and the environment temperature. In the temperature classification task, the positives were defined as the sample measurements higher than 36.2 ° C because it was difficult to obtain an abnormal body temperature in the preliminary experiment. The average temperature was used as a virtual threshold for triggering the temperature alarm.

4.1.5.3 Feature Design

This involved the design of a vector representation of each example, where the value of each vector element reflected an attribute of the example. The general principle of feature design is the construction of features that have good individual performance and are highly diverse. This is because the combination of complementary features tends to produce better performance than individual features. We designed 10 features from different aspects of the temperature signals, including the average temperatures for different time intervals and the environment temperature. Table 4-7 lists all the features used in this study. To clarify the notations, the first feature in Table 4-7, "Back, average temperature, 2s" indicates the average temperature over the 2s preceding the current sampling point as determined by the back sensor. "Chest, average temperature, 120s" indicates the average temperature over the 120s preceding the sampling point as determined by the chest sensor.

Table 4-7 Features used for temperature prediction and fever alarm

Feature ID	Feature description	Sensors ID
1	Back, average temperature, 2s	1
2	Back, average temperature, 60s	1
3	Back, average temperature, 120s	1
4	Chest, average temperature, 2s	2
5	Chest, average temperature, 60s	2
6	Chest, average temperature, 120s	2
7	Waist, average temperature, 2s	3
8	Waist, average temperature, 60s	3
9	Waist, average temperature, 120s	3
10	Environment temperature	4

4.1.5.4 Training

Using the annotated data, we were able to train a machine learning model to make predictions. For this purpose, we used the Waikato Environment for Knowledge Analysis (Weka) (Hall et al., 2009), a package that includes several machine learning models. For body temperature prediction, we tested many different regression models (Jang et al., 2007) including a linear regression model, least squares regression model, radius basic function (RBF) network regression model, Gaussian process model, and support vector regression model. We found that the Gaussian process model based on a kernel RBF had the best performance. For the temperature classification task, we tested the support vector machine (SVM) method using different kernels, as well as other methods such as logistic regression, Naive Bayes method and random forest method. We found that the performances of these classification methods varied, with the SVM method using a polynomial kernel being a little better. The detailed experimental results and analysis are presented in Section 4.1.6.

4.1.5.5 Testing

For both the tasks, we evaluated the performance using leave-one-out cross-validation strategy, where in each round one example was used as test data and the others as training, and the process was repeated until all the examples were tested. For the regression task, the evaluation measures were a correlation coefficient (CC) and mean absolute error (MAE), which are widely used in regression tasks. The measure is defined as Formula (4-1):

$$\text{CC (Correlation Coefficient)} = \frac{[\sum_{i=1}^{n} (X_i - \bar{X})(Y_i - \bar{Y})]^2}{\sum_{i=1}^{n}(X_i - \bar{X})^2 \sum_{i=1}^{n}(Y_i - \bar{Y})^2} \times 100\% \qquad (4\text{-}1)$$

where X_i and Y_i are the predicted value and the gold standard value, respectively, for the example; $\bar{X}$ and $\bar{Y}$ are their means. This metric ranges from 0 (the worst) to 1 (the best). The MAE is defined as Formula (4-2):

$$\text{MAE (Mean Average Error)} = \frac{1}{n}\sum_{i=1}^{n}|X_i - Y_i| \qquad (4\text{-}2)$$

This metric reflects the average discrepancy between the prediction and gold standard. A value of 0 indicates no error (perfect case). In the second task, we evaluated classification performance using some important measures of classification in machine learning, namely, Precision, Recall, F-score, Accuracy and AUC, as Formula (4-3) to Formula (4-5):

$$\text{Precision} = \frac{\text{\#true positives}}{\text{\#true positives+\#false positives}} \qquad \text{Recall} = \frac{\text{\#true positives}}{\text{\#true positives+\#false negatives}} \qquad (4\text{-}3)$$

$$\text{F-score} = \frac{2 \times \text{Precsion} \times \text{Recall}}{\text{Precision} + \text{Recall}} \qquad (4\text{-}4)$$

$$\text{Accuracy} = \frac{\text{\#true positives} + \text{\#true negatives}}{\text{\#all the examples}} \qquad (4\text{-}5)$$

Where "# true positives" is the number of correct predictions for the positive examples; "# true negatives" is the number of correct predictions for the negative examples; "# false positives" is the number of incorrect predictions for the positive examples; and "# false negatives" is the number of incorrect predictions for the negative examples. The F-score is the harmonic average of Precision and Recall; Accuracy reflects the ratio of the correct predictions; and AUC is the area under the ROC curve, insensitive to the threshold selection (Bishop & Nasrabadi, 2006).

4.1.6 Experimental Results

In the experiments, we observed the statistics and prediction performance of each individual sensor (Table 4-8 and Figure 4-10). As can be seen from the data, the temperatures determined by the skin-loose sensors were lower than the actual body temperatures measured by a thermometer and used as the gold standard. Moreover, the sensors were rather noisy, characterized by large ranges and variances. This is because the environment temperature was lower than the body temperature, and the signals measured by the sensors usually varied with the movement of the body. Predictably, the best individual sensor performance was rather poor, having a CC of 12.9% and MAE of 4.95, which made the measurements very unsuitable for practical use. The first three results in Table 4-9 were obtained by machine learning using real-time temperature measurements of individual sensors. Compared to the results in Table 4-8, the MAE of these latter results are significantly lower, although there is no significant improvement in the CCs. This indicates that machine learning scaled the signals to a similar range as the gold standard values (the reason for the reduced MAE), but learning from the real temperature signals received by an individual sensor was insufficient to obtain good fit with the actual body temperature (the reason for the poor CC).

Table 4-8 Performance of individual sensors based on real-time temperature

Sensors	Minimum	Maximum	Average	Deviation	CC	MAE
Back	28.69	33.38	31.22	0.93	7.16%	4.95
Chest	24.69	33.5	27.89	1.91	12.54%	7.57
Waist	26.5	33.5	29.78	1.41	12.90%	6.4

Note: CC: correlation coefficient; MAE: mean absolute error.

Table 4-9 Performance of body temperature prediction based on Gaussian processes mode

ID	Sensors	Machine Learning	Features	Correlation Coefficient	Mean Absolute Error
1	Back	No	None	7.16%	4.95 (Baseline)
2	Chest	No	None	12.54%	7.57
3	Waist	No	None	12.9% (Baseline)	6.4
4	Back	Yes	1	15.12%	0.24
5	Chest	Yes	4	11.29%	0.24
6	Waist	Yes	7	13.42%	0.23
7	Back	Yes	1,10	46.42%	0.21
8	Chest	Yes	4,10	33.78%	0.22
9	Waist	Yes	7,10	31.53%	0.22
10	Back	Yes	1,2,3,10	45.52%	0.21
11	Chest	Yes	4,5,6,10	38.01%	0.22
12	Waist	Yes	7,8,9,10	32.03%	0.22
13	Back+ Chest	Yes	1,2,3,4,5,6,10	50.99%	0.2
14	Back+ Waist	Yes	1,2,3,7,8,9,10	50.11%	0.2
15	Chest + Waist	Yes	4,5,6,7,8,9,10	40.95%	0.22
16	Back+ Chest+Waist	Yes	1,2,3,4,5,6,7,8,910	**53.91%(+317.9%)**	**0.2 (-96%)**

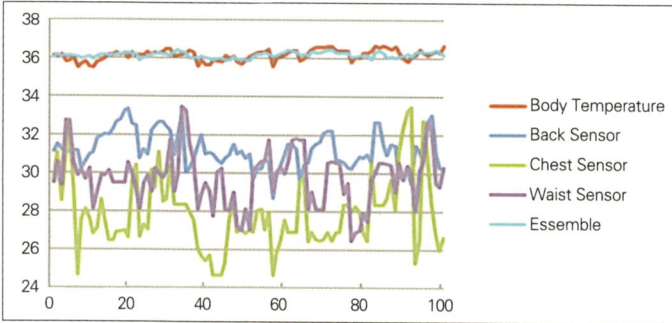

Figure 4-10 Real body temperature and the signals from the sensors

In Table 4-9, we compare the performances of the individual and combined sensors using the same machine learning framework. It is promising to see that the ensemble of different signals from multiple sensors (Run 16) significantly improves the prediction performance, achieving as much as 317.9% improvement of the CC and 96% reduction in MAE compared to methods without machine learning (Runs 1 – 3). There were also large improvements compared to the use of individual sensors with machine learning (Runs 4 – 6). From the changes between Runs 7 and 16, it can be concluded that the use of additional sensors or features tend to produce better results, confirming the advantage of machine learning methods for boosting weak signals. The incorporation of environment temperature considerably improved the CC (e.g. between Runs 4 and Runs 7). This is obvious from the fact that signals from skin-loose sensors were highly susceptible to environment temperature.

Considering that various machine learning algorithms have been proposed and utilized over the past few decades, we compared their results to identify the one most suitable for the present application. For the temperature prediction task, we compared six commonly used regression algorithms by implementing them in Weka, namely, linear ridge regression, least squares regression, RBF network, neural network, support vector machine and Gaussian process algorithms. The entire feature set in Table 4-8 was employed.

To ensure a fair comparison and avoid the possibility of over fitting, we used the default parameters of Weka for all the models, rather than tuning the parameters on the test set. It can be seen from Table 4-10 that the Gaussian process algorithm has the best performance. One possible reason for is that the signal follows a Gaussian distribution to some extent.

Table 4-10 Performance with different regression models

Model	Correlation Coefficient	Mean Absolute Error
Linear Regression	37.14%	0.22
Least Square	39.68%	0.23
RBF Network	13.42%	0.23
Neural network	23.06%	0.28
Support vector machine	42.92%	0.22
Gaussian Processes	**53.91%**	**0.2**

Table 4-11 presents the results of the temperature alarm task. Because the objective of the task was easier, i.e., to determine whether body temperature exceeded a threshold, the performances of all the algorithms were better compared to the determination of exact body temperature. From Table 4-10, it can be seen that the use of only the waist sensor without machine learning produced an accuracy of 66%, indicating that, although it is difficult to obtain an accurate

Table 4-11 Performance of high body temperature alarm using different methods

ID	Sensors	Machine Learning Model	Precision	Recall	F–score	Accuracy	AUC
1	Back	None	66.67%	30.43%	41.79%	61%	54.99%
2	Chest	None	51.69%	100%	68.15% (Baseline)	57%	53.58%
3	Waist	None	67.65%	50%	57.5%	66% (Baseline)	66.22% (Baseline)
4	Back+ Chest+Waist	Naive Bayes	70.8%	37%	48.6%	64%	66.2%
5	Back+ Chest+Waist	Logistic Regression	72.1%	67.4%	69.7%	73%	75.3%
6	Back+ Chest+Waist	Support Vector Machine (Poly – 3)	**73.5%**	**78.3%**	**75.8%** (+11.1%)	**77%** (+16.7%)	77.1%
7	Back+ Chest+Waist	Neural Network	68.3%	60.9%	64.4%	69%	73.1%
8	Back+ Chest+Waist	Random Forest	68.1%	69.6%	68.8%	71%	74.9%
9	Back+ Chest+Waist	Nearest Neighbor	69.2%	78.3%	73.5%	74%	74.3%
10	Back+ Chest+Waist	GMM	72%	78.26%	75%	76%	**77.15%** (16.5%)

estimate of absolute body temperature, improved performance can be achieved in the ranking of relative temperatures using a different threshold. For example, when two points were sampled, if the gold standard temperatures were 35.9 ° C and 36.5° C, respectively, and the corresponding measurements by the skin-loose sensors were 25.6 ° C and 29.5 ° C, the performance parameters of the temperature prediction (e.g., CC and MAE) would obviously be rather poor. However, if the purpose was to determine whether the measured temperature exceeded a set threshold of 27 ° C, the predicted temperature would still be considered correct. The SVM classification method using a degree 3 polynomial kernel and default regularization trade-off settings showed the best performance. It is also interesting to note that Run 10, using the results of Task 3, also produced a similar performance based on all the evaluation measures. This indicates that the temperature predictions are sufficiently general for application to a different task, thus eliminating the need to design another machine learning model for the second task. This simplifies the design of the proposed smart clothing.

We considered a different approach that combined multiple loose sensors located in different places in the clothes to get a better body temperature prediction. We used a regression model to find an approximation function between the multiple sensors' temperatures and the real body temperature. On the one hand, this method can improve the performance of skin-loose sensors, since it is well known that in using statistical regression method a complex function can be approximately represented as the combination of simple functions, that is, the combination of weak signals can be much stronger than the best individual signal. On the other hand, the comfort quality can be much higher than the skintight sensor setting, since it improves the accuracy without needing to use skintight sensors. In our experiment, we used three loose sensors placed on different parts of the body including the chest, back, and waist. The CC of the best individual sensors based on a single signal was approximately 12.9%. When we integrated various signals from all the sensors in a regression model, the performance increased to 53.9%. In practice, we usually need to know if the body temperature is over a certain threshold rather than predict the exact value; for example, we just need to measure whether body temperature is over 37 ° C. Therefore, we also designed another body temperature classification experiment that triggered an alarm if the body temperature rose above a certain threshold, and obtained 77% accuracy (Figure 4-11).

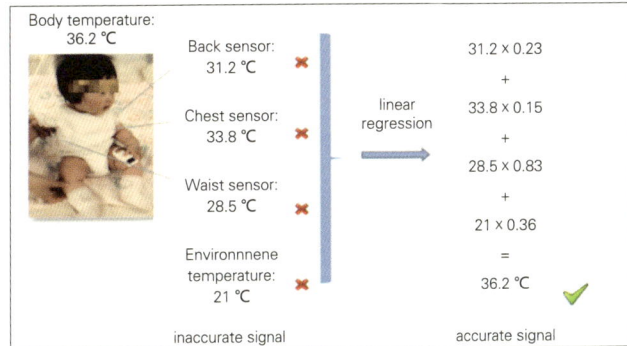

Figure 4-11 Performance enhancement via a linear combination of inaccurate sensors

The two tasks (body temperature regression and classification) were from two major aspects of supervised machine learning—regression and classification—the most active parts in modern artificial intelligence. We believe that the application of these techniques to smart clothing is essential because smart (intelligent) clothing can be viewed as a subfield of artificial intelligence. To our best knowledge, there is little research on the application of regression and classification methods to improve the comfort quality of smart clothing.

Finally, we record the process of collecting body temperature into the software by using Python program language. the software will be put on Cloud Server(DataV). Therefor, we could monitor infants' body temperature through computer or mobile phones at anywhere and anytime (Figure 4-12).

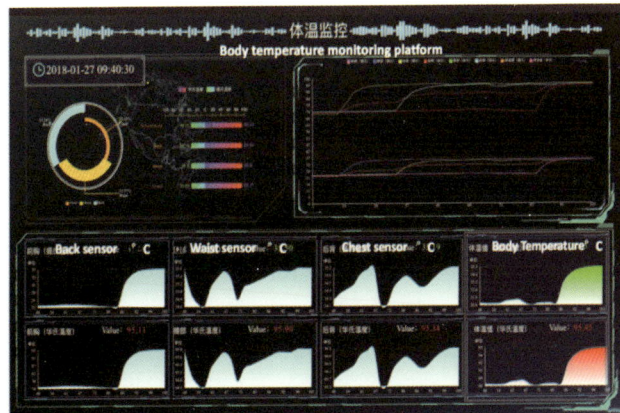

Figure 4-12 Data visualization LABV
source:https://datav.aliyun.com/share/7f55aaab9f4b8ec39a0f8da7964ae2c1.

4.1.7 Summary

We proposed a strategy for the design of smart clothing for monitoring body temperature using skin-loose sensors. Our experimental implementation and analysis of the strategy revealed that the combination of signals obtained by the employed multiple skin-loose sensors using a regression and classification model increased the accuracy of temperature measurement, while eliminating the discomfort caused by skintight sensors. This offers an effective means of overcoming the challenges faced in the utilization of smart clothing in daily life. Rather than the development of smart textiles, our study focused on the use of an artificial intelligence algorithm to improve the comfort of smart clothing without changing the hardware, namely, the textile and sensor type. The combination of state-of-the-art machine learning algorithms and simple pertinent features improved measurement accuracy between 10% and 300%, based on various important evaluation measures. This not only constitutes a significant achievement but encourages further study of the proposed strategy for smart clothing design, especially considering that some challenges remain, including the followings:

(1)Although skin-loose sensors are usually more comfortable than skintight sensors, the use of multiple sensors and the increased amount of wiring required create new challenges in terms of comfort and fashionability. One possible solution is the combination of smart textiles and machine learning to develop skin-loose textile-based sensors that can ease the discomfort of hard electronic sensors.

(2)The use of machine learning to develop smart clothing can be extended to several other applications of smart clothing, such as the monitoring of blood pressure, heart rate, voice and emotions. However, some of these tasks tend to be more difficult than monitoring body temperature because they require more complex and heavier sensors that are usually used in hospitals and difficult to incorporate in everyday clothing.

(3)The present study was conducted in a relatively simple environment, and it is not clear how well the proposed strategy would perform in a more complex situation, such as when applied to a person performing complex movements. This might require the design of a robust model that can handle such complexities, including data annotation and a machine learning algorithm.

(4)The "Design Plus X" thinking model facilities new collaborations between designers and engineers. Engineers were brought into design projects late in the process to provide behind-the-scenes, hidden technologies to support a design concept. "DPX" brings a fresh set of attitudes, aspirations, and capacities. It provides the expertise, sets new standards that others will rise to, and contributes to the development of capable and creative designers.

In the future, it would be interesting to see the expanded application of artificial intelligence to smart clothing for everyday monitoring of health using textile-based sensors, as well as the achievement of more effective modeling through the development of new methods for data annotation and evaluation. We plan to build framework of design of smart clothing for infants (Figure 4-13). Health-related information is gathered via wireless body-worn sensors and transmitted to the caregiver via an information gateway, such as a mobile phone. Caregivers can use this information to apply interventions as needed. The results of this study will connect with the meta-knowledge of design creativity (Nagai and Junaidy, 2015) to build a platform of design thinking and practice.

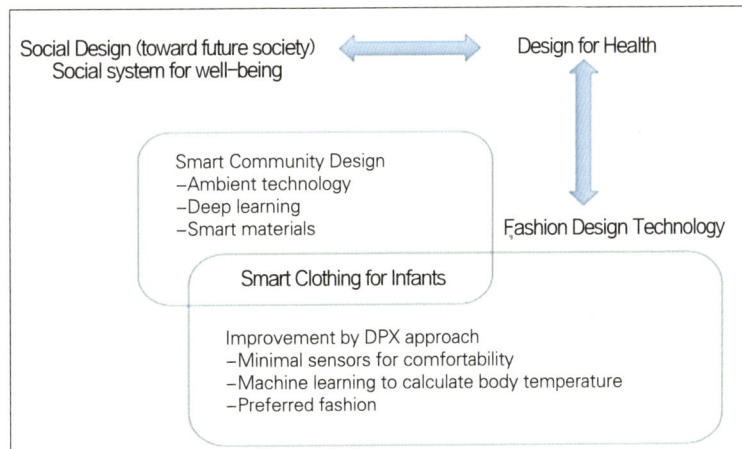

Figure 4-13 Framework of design of smart clothing for infants

4.2

Case Study 2

Project: Fashionable experience for blind children—design research of intelligent glove featuring perception of chromatic color .

Thinking Approach: Design Plus "X" .

Team members: DingWei (Fashion Design), Yukari Nagai (Knowledge Science), ZhangYunwei (Fashion Design), GouXiao (Information Science), XiaoLin (Information Science).

"DPX" workshop process:Figure 4-14.

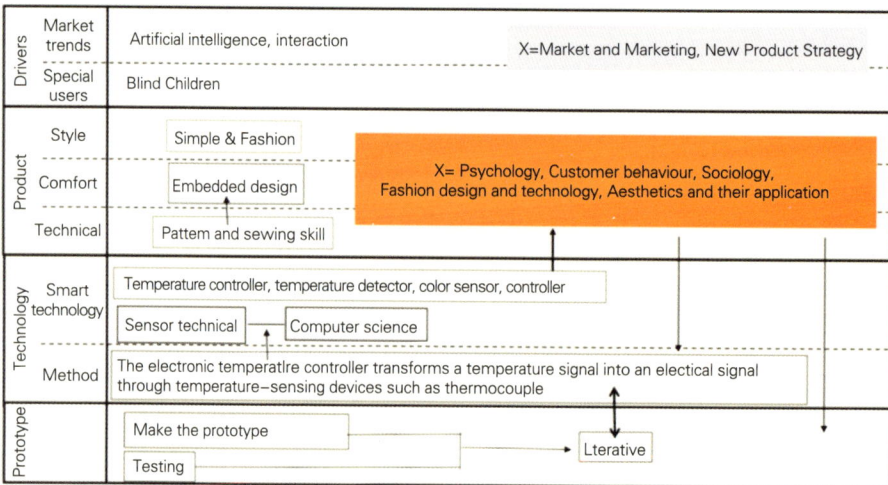

Drivers	Market trends	Artificial intelligence, interaction	X=Market and Marketing, New Product Strategy
	Special users	Blind Children	
Product	Style	Simple & Fashion	X= Psychology, Customer behaviour, Sociology, Fashion design and technology, Aesthetics and their application
	Comfort	Embedded design	
	Technical	Pattem and sewing skill	
Technology	Smart technology	Temperature controller, temperature detector, color sensor, controller / Sensor technical — Computer science	
	Method	The electronic temperatlre controller transforms a temperature signal into an electical signal through temperature–sensing devices such as thermocouple	
Prototype		Make the prototype / Testing — Lterative	

Figure 4-14 "DPX" workshop process

4.2.1 Research Introduction

Blind children are a special group of society. Loss of vision causes numerous obstacles for blind children in life and in study and makes the pleasure of reading, studying, and exploration of the world impossible for them. To ameliorate these limitations of blind children and improve their study experience, we have designed an intelligent glove that considers the handicaps of blind children and enables them to perceive and become aware of the existence of colors using temperature and chromatic-color sensors. When blind children who wear the intelligent glove touches a colored picture, the glove temperature changes with the change in the color coldness or warmness of the picture to perceive chromatic color and object shape. The intelligent glove helps blind children accurately "read" color illustration so that they can enjoy the pleasure of experiencing the picture and its accompanying text in a book. Meanwhile, attempt has been made to enable blind children to learn painting using this method and experience the full painting process. This intelligent glove provides a multi-element interactive experience mode to blind children so that they can better learn and perceive this colorful world through intelligent interaction.

4.2.2 Technical Support

4.2.2.1 Working Theory of the Sensor

Temperature controller, temperature detector, color sensor, controller, and other auxiliary electronic components are combined in the intelligent glove through embedding mode to make the glove capable of color perception. The color sensor (TCS 3200) statically recognizes the colors of an object, outputs different frequencies according to the color information, and transmits the output frequency to single-chip processor MC9S12XS128. In addition RGB color data are obtained after a frequency-sampling calculation using a single-chip processor. According to the RGB color table listed in annex A, a lookup table method is used to determine the color corresponding to its RGB value.

Figure 4-15 shows that numbers 1, 2 and 3 represent temperature controllers 1, 2 and 3 respectively. A temperature-detection sensor is simultaneously configured under each temperature controller, namely, temperature-detection sensors 1, 2 and 3. Temperature controller 1 is dependent on the control of the R value in RGB. The higher the R value is,

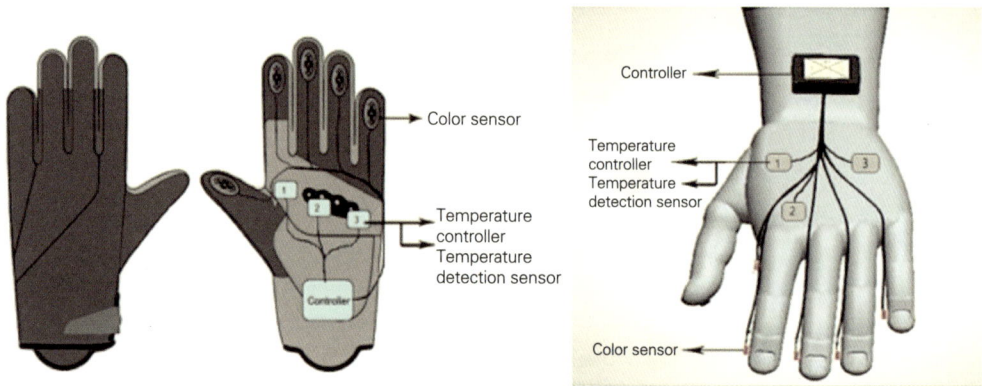

Figure 4-15 The numbers 1, 2 and 3 represent temperature controllers 1, 2 and 3 respectively

the higher is the temperature of temperature controller 1, that is, 0 – 255 respectively corresponds to 10~50 ° C. Similarly, temperature controller 2 depends on the control of the G value in RGB, and temperature controller 3 depends on the control of the B value in RGB. The temperature controller controls the temperature of the glove according to the statically recognized RGB color data by the color sensor (TCS 3200). Thus, the glove generates different temperatures and enables blind children to perceive color through temperature.

4.2.2.2 Working Theory of the Temperature Controller

The electronic temperature controller transforms a temperature signal into an electrical signal through temperature-sensing devices such as thermocouple and platinum resistor and controls a relay to activate (or deactivate) heating (or refrigerating) equipment using circuits such as single-chip processors, programmable logic controller, and other devices. For the purpose of this study, the temperature-control system of the intelligent glove utilizes temperature sensor DS18B20 and features an intelligent temperature-controller design. In this system, the DS18B20 three-core mini-type contact-chip temperature sensor performs complete environmental temperature signal collection and transmits the collected signal to a single-chip processor (MC9S12XS128) for processing and complete intelligent control.

4.2.2.3 Working Theory of a Color Sensor

Color sensor is also referred to as chromatic-color recognition sensor. The color sensor is designed to consist of an independent photodiode with corrected red, green, and blue optical filters and to perform relevant processing of the output signal so that the color signal is recognized. The basic theory of color recognition is described as follows.

(1)Color characteristics.

①Hue is based on the wavelength and is used to distinguish the characteristics of different colors.

② Saturation reflects the purity of color; each color type can be understood as the result of the mixture between a certain type of spectral color and a white color. The higher the proportion of the spectral color is, the higher is the color saturation, and vice versa.

③ Lightness describes an attribute of the color brightness and is related to light energy as a measure of light intensity.

(2)Trichromatic theory.

We appropriately select three primary colors (red, green, and blue), and combine these colors according to different proportions to generate a visual sense of the different colors. The overall lightness of the three primary colors determines the lightness of the combined color lights. The proportion of the three primary color components determines the chrominance. The three primary colors independently exist, and making up any primary color is impossible.

(3)Semiconductor characteristics.

Conductivity substantially changes depending on the external optical and thermal stimuli, namely, light- and thermo-sensitive elements. The basic steps of color recognition are as follows: using a color sensor (mainly a color-sensitive sensor) to transform optical signals into an electrical current; pretreatment of the electric current micro signal, analog to digital conversion, and sending the digital signal to a single-chip processor or a microcomputer for processing.Theory of recognition: color sensor detects color by comparing the object color with a standard reference color and outputting the detection results when the color of an object matches the standard color within a certain range of error.

4.2.2.4 Technical Support for Embedding Design

Intelligent wearable devices usually select an embedding method for integration of hardware equipment such as sensor and electronic components. This hardware equipment is embedded into intelligent wearable devices without affecting the comfort of a person using such wearable devices and in such a manner that it is difficult for the user to be aware of the device. Therefore, hardware devices are excellently integrated into intelligent wearable devices. This design adopts the embedding design technique so that sensing function of the intelligent gloves can be ensured, and wearing comfort and good appearance can be realized, as shown in Figure 4-16.

4.2.3 Experimental Design

4.2.3.1 Relationship Diagram of the Main Components of the System

The power supply module provides 5- and 12-V power supplies to the single-chip processor and sensor, respectively. When the color sensor detects a color signal, the color signal is converted into a specific electrical signal and inputted into the single-chip processor. The single-chip processor determines the temperature corresponding to the color using a lookup table method according to the data transmitted from the color sensor and then sends a warm-up or cool-down command to the temperature controller. Becoming aware of the temperature rise or drop caused by the temperature controller is impossible for the single-chip processor itself. Thus, a

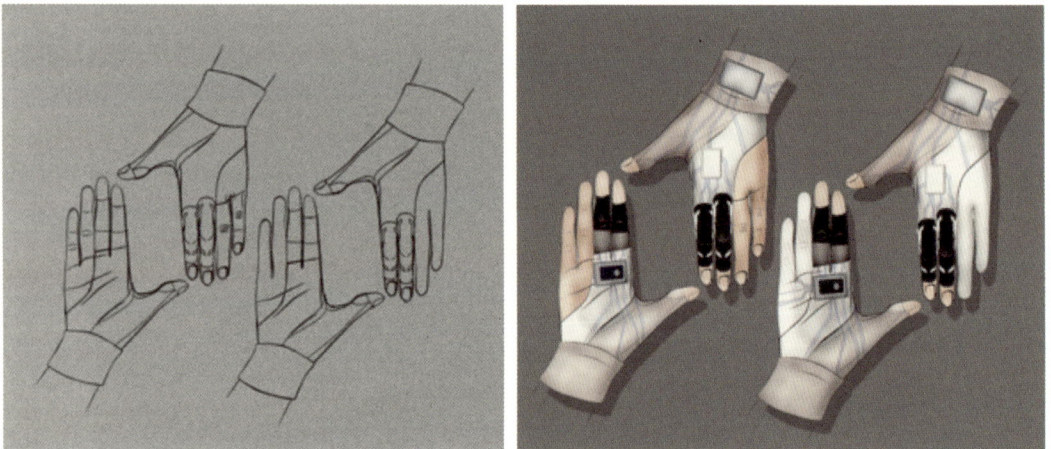

Figure 4-16 The illustration of smart gloves

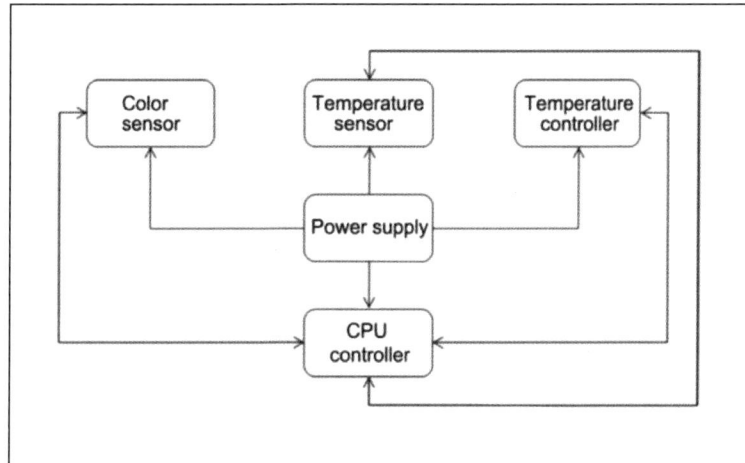

Figure 4-17 Relationship diagram of the main system components

temperature sensor is needed to monitor the temperature in real time, which is then fed back to a single-chip processor (Figure 4-17).

4.2.3.2 Specific Control Procedure Flow

(1)Color sensor (TCS 3200) is used to detect data corresponding to the color identified by the current sensor.

(2)The color signal is transmitted to the single-chip processor, which calculates the color corresponding to the current data using a formula.

(3)A liquid-crystal display (LCD) screen directly displays the current color in word form (to allow a person with a normal vision to check the operating status of the sensor at any time).

(4)The lookup table method is utilized to determine the temperature to be controlled corresponding to the current color.

(5)The control mode of a proportional—integral—derivative controller is utilized to manage the temperature-control sensor (i.e. refrigeration element) for heating or refrigeration.

(6)A temperature-detection sensor is employed to detect the current temperature of the temperature-control sensor (i.e. refrigeration element) in real time and to feed back the signal to the single-chip processor.

(7)The single-chip processor determines the real-time operating status of the temperature-control sensor (i.e. refrigeration element) using a comparison method.

(8)A digital tube displays the current temperature of the temperature-control sensor (i.e., refrigeration element) in real time to allow a person with normal vision to check the operating status of the sensor at any time.

4.2.3.3 Process Flowchart (Figure 4−18)

4.2.3.4 Temperature Setup

(1)Experiment on hand discrimination of temperature difference.

A cold or warm color refers to the temperature difference in the chromatic color. For chromatic purpose, colors are divided into warm color (red, orange, and yellow), cool color (cyan and blue), and neutral color (violet, green, black, gray, and white) according to psychological sensation.

Three temperature ranges are obtained through experimental temperature measurement: cool color: 10~20° C, neutral color: 20~30° C, and warm color: 30~40° C. The discrimination by hand of the temperature difference is detected in these three ranges.

In the 10~20 ° C temperature range listed in Table 4-12, the discrimination by hand of the temperature is recognizable for a temperature difference of 2° C, those for temperature differences of 3° C and 4° C are both high. In the 0~20° C temperature range, we can select a temperature difference of 2~4° C to design cool-color temperature.

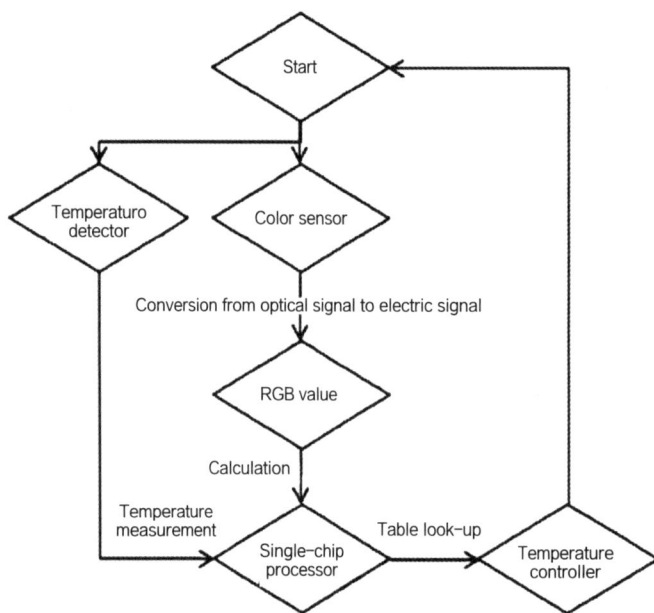

Figure 4-18 Process flowchart

Table 4-12 Experiment of hand discrimination of the temperature difference

Temperature difference	10~20°C	20~30°C	30~40°C
2°C	Recognizable	Recognizable	Recognizable
3°C	Recognizable	Highly recognizable	Highly recognizable
4°C	Highly recognizable	Highly recognizable	Highly recognizable

In the temperature range of 20~30° C, the discrimination by hand of the temperature is recognizable at a temperature difference of 2° C. Those for temperature differences of 3° C and 4° C are both high. In the temperature range of 10~20° C, we can select a temperature difference of 2~4° C to design neutral-color temperature.

In the temperature range of 30~40° C, the discrimination by hand of the temperature is recognizable at a temperature difference of 2° C. Those for temperature differences of 3° C and 4° C are high. In the temperature range of 10~20° C, we can select a temperature difference of 2~4° C to design a warm-color temperature.

The color temperature of the temperature controller (degree of coldness or warmness) is controlled by the RGB value, and different RGB colors are obtained from the changes in the three color channels of red, green, and blue as well as their mutual superposition. RGB colors represent the three channels of red, green, and blue. The RGB colors are set up according to the theory of color luminescence. For better understanding of this concept, it is similar to the lights of three lamps with red, green, and blue colors superimposed among one another. Chromatic colors are mixed, and the color lightness represents the overall superimposed lightness. The more intense the color overlap is, the higher is the color lightness, i.e., the sum of the mixing characteristics. The more intense the superposition is, the higher is the lightness. Each color of the three color channels is divided into 255 orders of lightness. A lamp lightness of zero indicates the weakest lightness, whereas the lamp lightness is highest at 255. When the values of the three colors are identical, it appears gray. It is a bright white when all three colors are at 255, and it is black when all colors are zero. Therefore, the temperature controller can control the temperature according to the

RGB chromatic colors. The temperature of the temperature controller is higher when the R value in the RGB color is larger, and it is lower when the B value is larger. G influences R and B and affects the cool-down function for R and warm-up function for B. Thus, temperature control realizes control of the color temperature in accordance with the above theory as per the RGB value.

(2)RGB cold and warm color relationship and temperature setting range.

In Table 4-13, the higher the R value is, the higher is the temperature. The higher the B value is, the lower is the temperature. The G value influences the R and B values. The larger the G value is, the larger is the influence on the R and B values. When R value > B value, the color is a warm color. The larger the difference is, the higher is the temperature. The G value plays the role of cooling down at this point. The larger the G value is, the larger is the temperature drop. The range of colors is red, orange, and yellow, and the temperature setup range is 30~40° C. When R value < B value, the color is a cool color. The larger the difference is, the lower is the temperature. The G value plays the role of warming up at this point. The larger the G value is, the larger is the temperature rise. The range of the colors is cyan and blue, and the temperature setup range is 0~20° C. When R value = B value, the color is neutral, and the G value has no effect. The range of the colors is violet, green, black, gray and white, and the temperature setup range is 20~30° C in Table 4-13.

(3)Color temperature setup experimental results is in Table 4-14.

Table 4-13 RGB cold and warm color relationship and temperature setting range table

Relationship between R and B values	Cold/warm color	Role of G value	Range of colors	Temperature setting range
R value > B value	Warm color	Cool–down	Red, orange and yellow	30~40° C
R value < B value	Cool color	Warm–up	Cyan and blue	10~20° C
R value = B value	Neutral color	No effect	Violet, green, black, grey and white	20~30° C

Note: R value = B value = G value = 255 is white. R value = B value = G value = 0 is black. otherwise, all other combinations result in neutral color.

Table 4-14 Color temperature setup experimental results

Serial No.	Color	RGB valuc	Temperature	Temperature difference	Degree of coldness or warmncss	Degree of comfort	Discrimination
1	Red	255,0,0	40° C	4° C	Hot	Comfortable	High
2	Pink	219,112,147	28° C	4° C	Warm	Comfortable	Recognizable
3	Salmon pink	255,69,0	36° C	3° C	Moderately hot	Comfortable	High
4	Orange	255,165,0	33° C	3° C	Very warm	Comfortable	High
5	Yellow	255,255,0	30° C	3° C	Warm	Comfortable	High
6	Olivine	154,205,50	24° C	2° C	Slightly warm	Comfortable	Recognizable
7	Forest green	34,139,34	22° C	2° C	Moderately cold	Comfortable	Recognizable
8	Lemon green	0,255,0	20° C	2° C	Cool	Comfortable	Recognizable
9	Sky blue	0,191,255	16° C	3° C	Cold	Comfortable	Recognizable
10	Dark blue	0,0,139	10° C	3° C	Extremcly cold	Uncomfortable	High
11	Blue	0,0,255	13° C	3° C	Cold	Uncomfortable	High
12	Purple	128,0,128	18° C	2° C	Cold	Comfortable	Recognizable

Note: R value = B value = G value = 255 is white. R value = B value = G value = 0 is black. otherwise, all other combinations result in neutral color.

4.2.4 Color- and Temperature-recognition Experiments

This system has a very high requirement on the relevance between the software and hardware, and the change process of the entire experiment is very complicated because everything is subject to the influence of time, space, and other numerous factors; avoiding certain slight errors is impossible.

The specific operation of color- and temperature-recognition experiments in this study is described as follows: First, the power supply of the development board is turned on, and a chromatic color cardboard is flatly

Table 4-15 Color- and temperature-recognition experimental data table

Color	Result		RGB	RGB Value	Output RGB	Temperature (°C)	Temperature Identification	Photo
Red		Exper iment1	255, 0, 0	√	170, 36, 47	40° C	√	
		Exper iment2		√	170, 34, 46	40° C	√	
		Exper iment3		√	170, 36, 43	40° C	√	
Pink		Exper iment1	219, 112, 147	√	201, 84, 125	28° C	√	
		Exper iment2		√	203, 86, 130	28° C	√	
		Exper iment3		√	201, 84, 127	28° C	√	
Salmon pink		Exper iment1	255, 69, 0	√	190, 58, 25	36° C	√	
		Exper iment2		√	190, 60, 26	36° C	√	
		Exper iment3		√	192, 57, 22	36° C	√	
Orange		Exper iment1	255, 165, 0	√	223, 139, 10	33° C	√	
		Exper iment2		√	233, 137, 9	33° C	√	
		Exper iment3		√	222, 136, 12	33° C	√	
Yellow		Exper iment1	255, 255, 0	√	209, 199, 23	30° C	√	
		Exper iment2		√	209, 195, 30	30° C	√	
		Exper iment3		√	208, 196, 31	30° C	√	
Olivine		Exper iment1	154, 205, 50	√	118, 156, 37	24° C	√	
		Exper iment2		√	119, 157, 39	24° C	√	
		Exper iment3		√	117, 158, 40	24° C	√	
Forest green		Exper iment1	34, 139, 34	√	56, 109, 49	22° C	√	
		Exper iment2		√	55, 110, 46	22° C	√	
		Exper iment3		√	55, 108, 47	22° C	√	

Note: There are certain slight errors in the experimental process.

placed. Then, a color-collection module under design is rightly and flatly placed on a cardboard under test. The color-collection module is covered by a paper with good light-isolation performance. The value shown in the LCD is then recorded using a pen. Boards of different colors are measured in turn, and the measured data are recorded. In the experiment, three tests are performed for each color. The recorded data are listed in Table 4-15.

4.2.5 Analysis of the Experimental Results

As presented in the above-mentioned experimental results, the intelligent glove that features perception of a chromatic color can generate 12 different temperatures by recognizing 12 colors. Hence, a user can perceive different colors. This design has a high requirement on the relevance between the software and hardware. The program design is complicated, and the fabrication is very difficult. Therefore, certain slight errors could be generated in the experimental process, but these will not affect the intelligent glove feature perception of chromatic colors in terms of recognizing the 12 colors and temperature control. Certain errors occur in the color displayed by the LCD panel and the setup standard color. Numerous factors may generate these errors, such as the existence of external interfering light, different sensitivities to light of the sensor chips, instability of the light emitted from the LED diode in the light supplement module and certain other influential factors. Therefore, certain errors will unavoidably exist in the experimental measurement, but these do not affect the intelligent glove in terms of recognition of the 12 colors and temperature control.

4.2.6 Summary

This research attempts to combine art with science and technology by combining advanced sensing technology with art painting elements to design an intelligent product, intelligent glove that features chromatic-color perception.This intelligent product design of an intelligent glove provides multi-element painting experience to people who are enthusiastic about art and painting and enables special group of blind persons to experience painting.

As a novel model of intelligent wearable device designed for blind persons to perceive chromatic colors, this product combines intelligent technology with painting skill and chromatic-color attributes. The innovation point of the design is to enable blind persons to perform actions similar to a person with good eyesight, experience chromatic colors, perform painting in a different manner, and utilize different painting experience and painting skills in the painting education and reading fields for blind children, which can enable them to enjoy the pleasure of painting and reading.We have broken a natural limitation. We possibly can not only observe things through our eyes but also perceive things through temperature through the painting process. A chromatic-color element is combined with a temperature element. The temperature is used to perceive chromatic color, the chromatic-color is sufficiently utilized, and better perception and expression of things are realized. Thus, blind persons can be brought out from the dark world and be enabled to perceive the multicolored world just like a normal person.

The design in this paper conveys and embodies not only an intelligent wearable device but also care for special groups of disabled person as well as appreciation and respect for life.

Chapter 4 contains two case studies, one project is" Designing Comfortable Smart Clothing: for Infants' Health", the other one is "Fashionable Experience for Blind Children—Design Research of Intelligent Glove Featuring Perception of Chromatic Color". This chapter documents the experimental stage of the practical research and explains the two case studies employed the DPX thinking approach to create the new production and got the good design outlets.

Cases study 1:

(1)Journal paper.

Ding Wei, Yukari Nagai, Liu Jing, Guoxiao. Designing Comfortable Smart Clothing: For Infant's Health Monitoring, [J].International Journal of Design Creativity and Innovation, DOI:10.1080/21650349. 2018.1428690.

(2)Oral presentation at international conference.

Ding Wei, Yukari Nagai, Liu Jing. Smart Clothing Design: A Machine Learning Approach, 2-5th November, IASDR 2015 interplay, Brisbane, Australia.

(3)Poster presentation at international conference.

Ding Wei, Yukari Nagai, Liu Jing. Intelligent Reactions Between Human Body and the Environment: Design of Smart Clothing. 2-5th November, IASDR 2015 interplay, Brisbane, Australia.

(4)Poster presentation at international symposium (best poster award).

Ding Wei, Yukari Nagai, Liu Jing. Intelligent reactions between human body and the environment.26-29th March 2016, JAIST HLD Symposium, Kanazawa, Japan.

Cases study 2:

Oral presentation at international conference.

Yukari Nagai, Ding Wei. Intelligent Glove Featuring Perception of Chromatic Color. 4-7th December 2017, Design 4 Health, Melbourne, Australia.

Chapter 5

Discussions and Conclusions

5.1

Research Discussions

(1)Social needs drive design education.

What is the driving force behind design education? (Figure 5-1) The context in which we live exerts a decisive influence on the nature of education and it determines the meaning of what it is to be educated. The profession for which we educate designers today takes place against a context with several dimensions. From the 18th to 19th Century, "Conventional Design" focuses on linear design thinking, the quality of the product itself, and physical design. From the 19th to 20th Century, "Design Thinking" focus on user-centered, innovative design, experience design, and satisfy customer innovation needs with empathy. After the 20th Century, with the rapidly developing technology, multidimensional social issues, design transform into the information, knowledge processes, programming, algorithms, computers, data transmission and human interaction of post-industrial times. Conventional design education can't meet the requirements of social development. Our research work try to establish the new thinking approach "Design Plus X" to meet the needs of the society to further the development of design human resources needs. "DPX" focus on re-establishes the relationship between people and the world and human well-being.

The necessity for highly specialized human talent that has newly emerged due to technological, economical, and societal developments is the driving force propelling changes in higher education in fashion, while at the same time indicating the direction of reform and development within fashion design education. Fashion design education should be divided into three parts (Figure5-2):

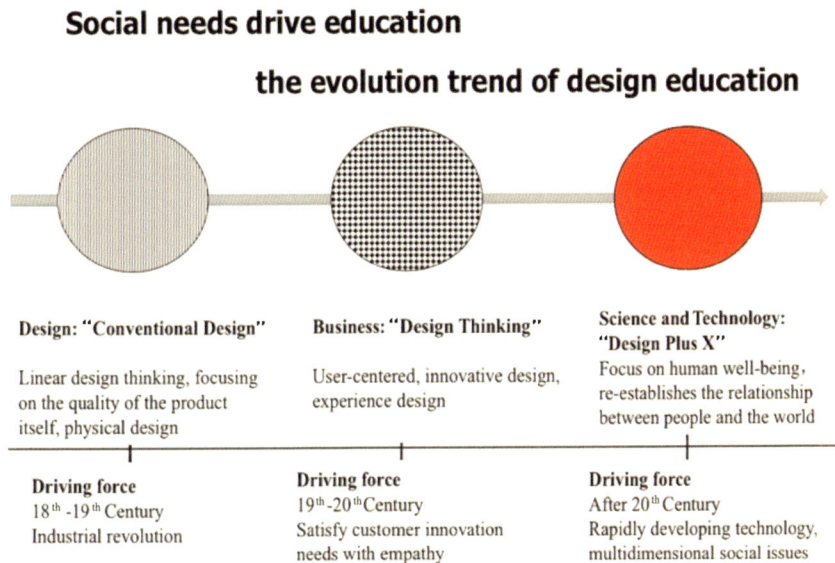

Social needs drive education

the evolution trend of design education

Design: "Conventional Design"

Linear design thinking, focusing on the quality of the product itself, physical design

Business: "Design Thinking"

User-centered, innovative design, experience design

Science and Technology: "Design Plus X"

Focus on human well-being, re-establishes the relationship between people and the world

Driving force
18ᵗʰ -19ᵗʰ Century
Industrial revolution

Driving force
19ᵗʰ -20ᵗʰ Century
Satisfy customer innovation needs with empathy

Driving force
After 20ᵗʰ Century
Rapidly developing technology, multidimensional social issues

Figure 5-1 Social needs drive education/the evolution trend of design education

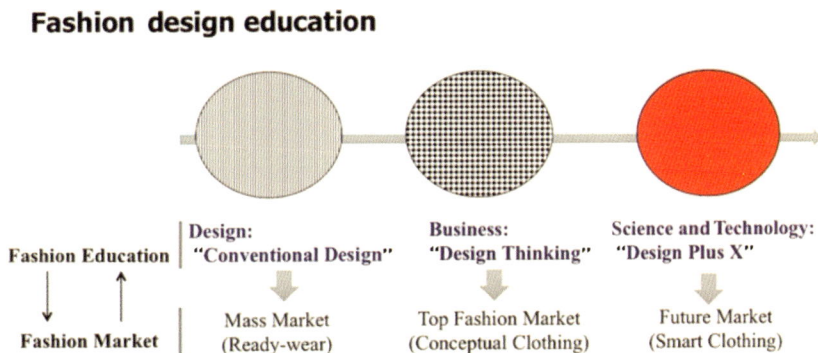

Fashion design education

| | **Design:** "Conventional Design" | **Business:** "Design Thinking" | **Science and Technology:** "Design Plus X" |

Fashion Education

Fashion Market

Mass Market (Ready-wear)

Top Fashion Market (Conceptual Clothing)

Future Market (Smart Clothing)

Figure 5-2 The fashion design education corresponds to the fashion market

①"Conventional Design" corresponds to the mass market.

② "Design Thinking" corresponds to the Top fashion (conceptual clothing).

③ "DPX " corresponds to the future market (smart clothing). We believe smart clothing can provide more function to people. It will build a new relationship between clothing and human. Designing functional smart clothing will help or support people to get more ability. Further, smart clothing can provide a new solution for social issues, and contribute to the higher quality of human life.

The necessity for highly specialized human talent that has newly emerged due to technological, economical and societal developments is the driving force propelling changes in higher education in fashion, while at the same time indicating the direction of reform and development within fashion design education. Fashion design education should be divided into three parts (Figure 5-2):

①"Conventional Design" corresponds to the mass market.

②"Design Thinking"corresponds to the top fashion (conceptual clothing).

③"DPX" corresponds to the future market (smart clothing).

We believe smart clothing can provide more function to people. It will build new relationship between clothing and human. The functional smart clothing design will help or support people to get more abilities. Further more, smart clothing can provide new solutions for social issues, and contribute to higher quality of human life.

(2)All forms of professional design education can be assumed to be developing design thinking in students, even if only implicitly.

What is the difference between "Design Plus X" and the Design Thinking approach and UX/UXD? (Table5-1). Based on research and work experience the author created the term of "Design Plus X"."DPX" through the integration of multiple domains (interdisciplinary path) is encouraged

Table 5-1 A comparison of three models of creativity

Thinking Model/Approach	The Logic of Thinking
Design thinking	Through creative strategies to solve the problem. Use empathy to solve problem
UX/UXD	Human−centered experience design. Solve the problem of User needs
Design plus X	Through the integration of multiple domains (interdisciplinary path) is encouraged in order to respond to the need for societal well-being. Facilities new collaborations between designers and engineers

in order to respond to the need for societal well-being. The "Design Plus X" thinking model can facilitate new collaborations between designers and engineers. Engineers were brought into design projects late in the process to provide behind-the-scenes, hidden technologies to support a design concept. "DPX" brings a fresh set of attitudes, aspirations and capacities. It provides the expertise and sets new standards that others will rise to, and contributes to the development of capable and creative designers.

Here are a few things about DPX and Design Thinking:

①It insists keep focused on human well-being and social issues.

②Interdisciplinary approach.

③Aligning with science and technology.

④It relies on both creativity, experiment test and logic.

⑤Teamwork.

⑥Reestablishing the relationship between people and the world.

(3)DPX: strategies for full integration achievement for designers and engineers.

Smart clothing was developed to support innovative decisions made in the sourcing and selection of materials, technologies, and construction methods. The process includes identifying end-user needs, fiber and fabric development and textile assembly, and garment development. To balance appearance and function, designers require guidance in their selection and application of technical textiles, style, cutting, sewing and finishing at every stage in the design research and development process. Yet intrinsic factors may be more important than the results of extrinsic factors when considered from the viewpoint that the design thought process is the designer's creative expression process (Yukari Nagai, Candy & Edmonds,2007). For the project of Smart Clothing Design, it was important to develop a collaboration basis. How to balance all the contributions, DPX thinking approach considered the intrinsic factors between designers and engineers, and built the strategies for integration achievement. It contained a clear goal and expected contributions from all participant (Figure 5-3).

(4)Setting up a "DPX Studio" practical learning site, wechat and blogs.

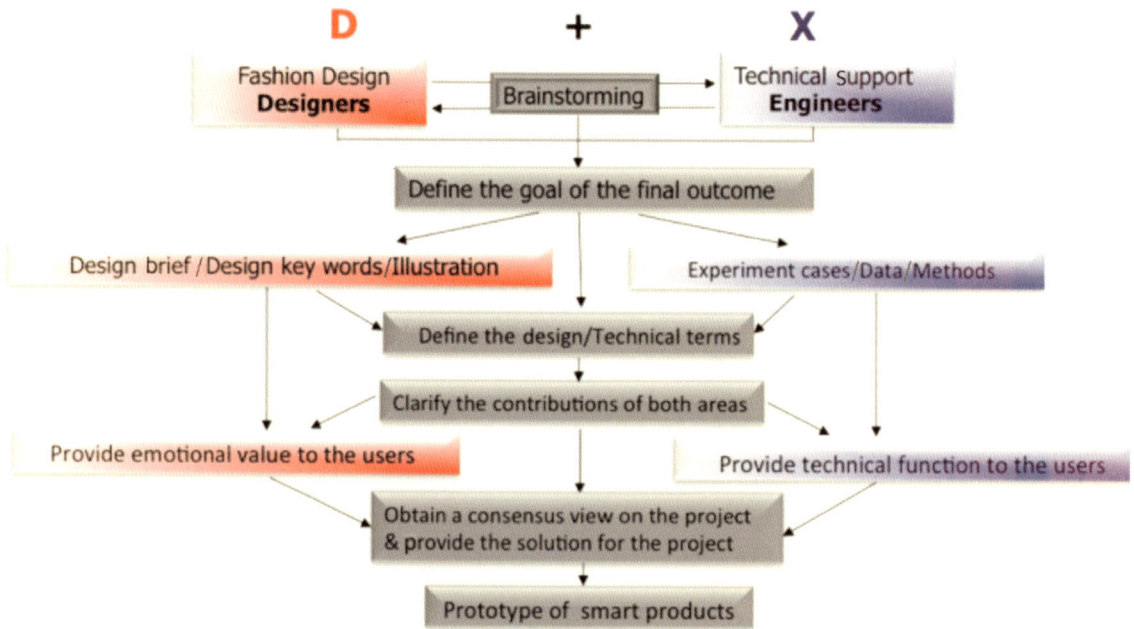

Figure 5-3 DPX: strategies for full integration achievement for designers and engineers

Multimedia will be fully utilized, and online sharing of educational resources will be realized. A learning site will be constructed and a web platform aiming for new reform and exploration of practical learning will be established, serving as a medium and intermediary of practical learning. With a focus on the construction of a database module, work item module, and an interactive communication module, students will effectively exercise and organize all educational activities while implementing DPX. WeChat and blogs will be set up, and students will be able to discuss the results gained from practical learning through the learning site, and share their experiences.

For now, DPX studio totally have 370 students:76% students are from fashion school,15% students are from information science school, 7% students are from management school, 2% students are from others (Figure 5-4).From June 2017 to now, the workshop has completed 10 projects successively (Figure5-5, Figure5-6).

(5)Construction of an international DPX research platform.

Through the practical education project platform, an international educational research and departmental research platform for teaching staff and students of higher education institutions will be constructed. International project coordination and exchange will be strengthened,

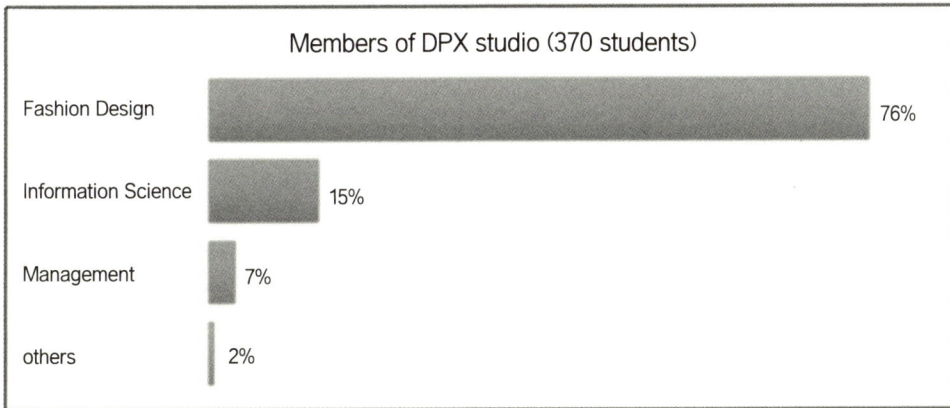

Figure 5-4 The proportion of students in the DPX studio

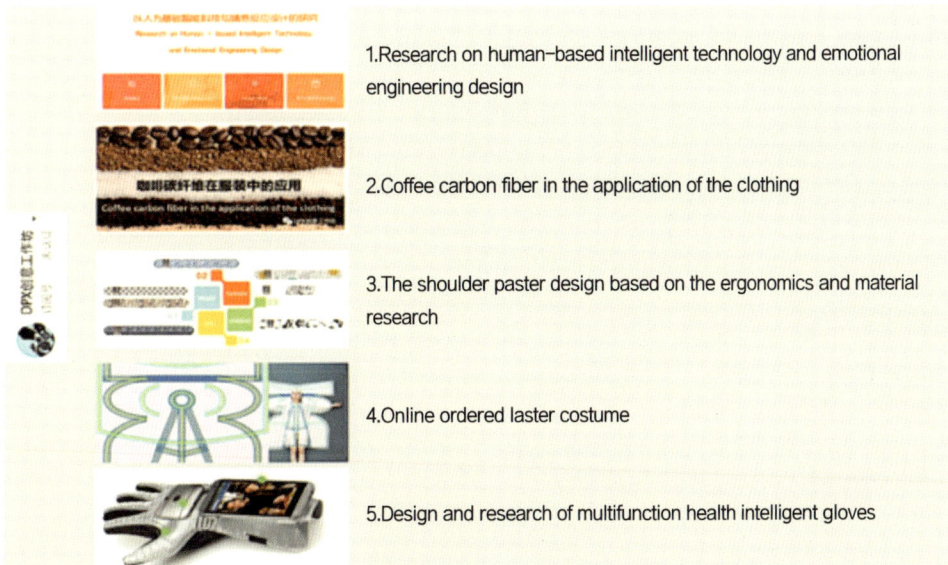

1.Research on human-based intelligent technology and emotional engineering design

2.Coffee carbon fiber in the application of the clothing

3.The shoulder paster design based on the ergonomics and material research

4.Online ordered laster costume

5.Design and research of multifunction health intelligent gloves

Figure 5-5 The case (1-5) studies in the DPX studio

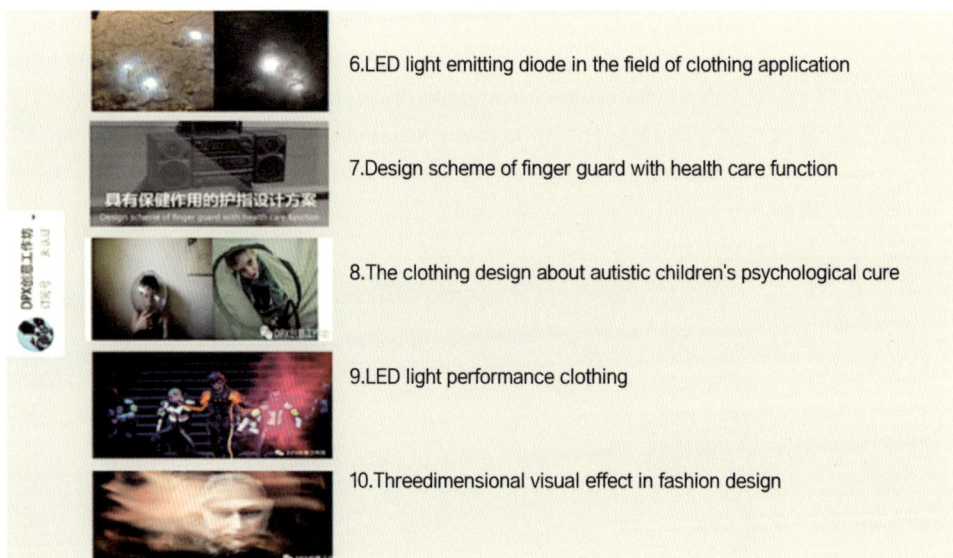

6.LED light emitting diode in the field of clothing application

7.Design scheme of finger guard with health care function

8.The clothing design about autistic children's psychological cure

9.LED light performance clothing

10.Threedimensional visual effect in fashion design

Figure 5-6 The case (6-10) studies in the DPX studio

the outlook of teaching staff will be broadened via academic exchange and research debate, and research at the cutting edge of manufacture will be facilitated. Furthermore, corporate leaders must be added to the ranks of supervisors in order for students to learn from market experts. Efforts will be made towards the construction of a top-level industry-university cooperation model that transcends the conventional single cooperation model in which individual members of teaching staff and corporations carry out one single project, and research resources that can be mobilized will be vastly increased.

(6)Interdisciplinary practice is useful for fashion designers?

Regarding whether interdisciplinary practice is useful for fashion designers, we did a questionnaire survey. The 410 FD students surveyed, 93.5% generally agreed on the significance of interdisciplinary practical (Figure 5-7).

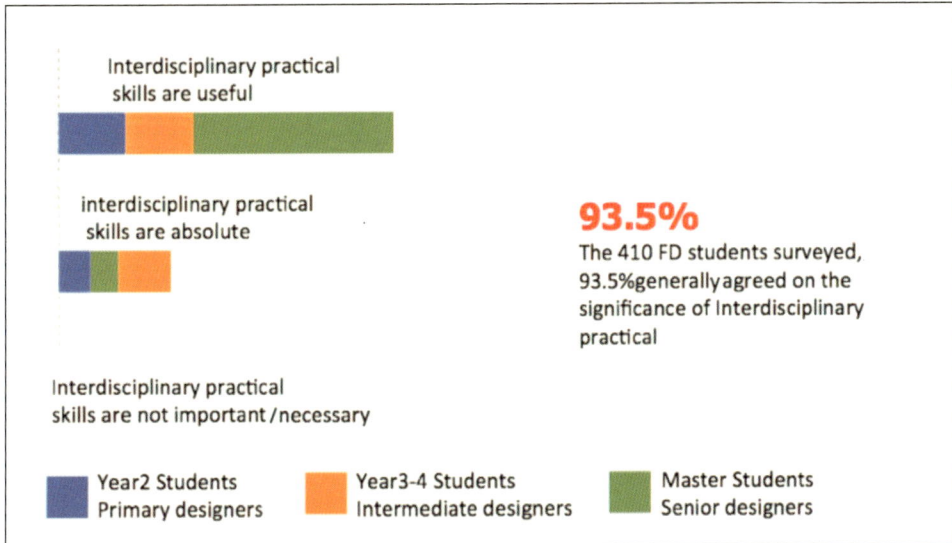

Interdisciplinary practical
skills are useful

interdisciplinary practical
skills are absolute

93.5%

The 410 FD students surveyed,
93.5% generally agreed on the
significance of Interdisciplinary
practical

Interdisciplinary practical
skills are not important /necessary

Year2 Students
Primary designers

Year3-4 Students
Intermediate designers

Master Students
Senior designers

Figure 5-7 The questionnaire results about the interdisciplinary practice in fashion education

5.2

Research Conclusions

5.2.1 Research Contributions

This research has two key contributions.

5.2.1.1 Research Contribution 1

Firstly, the research proposed a new strategy for interdisciplinary practical in Fashion education called "Design Plus X" (DPX), as an integrated new design thinking approach based on multidiscipline crossover.

"X" corresponds to knowledge from various crossover departments and fields (e.g. sociology, management science, electronic engineering, information science, etc). The purpose of the new practical system research called Art Design DPX, with the whole process of design development as a carrier, to incorporate the knowledge from various crossover departments and fields represented by X into design, utilize the broadening and deepening of knowledge related to design development, to comprehensively crossover multidisciplinary fields such as Design Plus Humanities and Sociology, Design Plus Management and Design Plus Technology, and to develop an integrated, innovative practical education. This new thinking approach Design Plus X (DPX), has several advantages, summarized as follows:

(1)New strategy.

The "Design Plus X" strategical thinking approach facilitates new strategy for interdisciplinary practical fashion education, it offered an integrated new design thinking approach based on multidiscipline crossover.

(2)New integrated concept.

"DPX" with the whole process of design research development as a carrier, multidisciplinary fields will crossover comprehensively, and by enforcing the consistency of Design Plus X, a new integrated concept for design practice within practical education, overemphasis on formal reorganization in design education can be resolved, and a core capacity for innovation that centers on reform of method and function will be cultivated.

(3)New standards.

Based on DPX thinking approach, it brings a fresh set of attitudes, aspirations, and capacities. It provides the expertise, sets new standards that others will rise to, and contributes to the development of capable and creative designers. It is encourage designer from different way to thinking problem, mix knowledge break discipline bound to solve issues of fashion design field. The integration of multiple domains is encouraged in order to respond to the need for societal wellbeing.

(4)New perspective.

We apply the KS to support smart clothing design research work from designer perspective, promote research and education in the field of Knowledge Science based on Fashion Design and Smart Clothing Design. The research works will contribution to develop KS into design for human life.

5.2.1.2 Research Contribution 2

Secondly, DPX for humanities-oriented research, conception of commercial strategies and plans for technological realization will be incorporated into practical education, and practical training about integrated solutions related to product design and services will be conducted.

This thesis contains two case studies: One project is" Designing Comfortable Smart Clothing: for Infants' Health". The other one is "Fashionable Experience for Blind Children—Design Research of Intelligent Glove Featuring Perception of Chromatic Color". The thesis has made many practical researches in the experimental stage and explained the two case studies with the DPX thinking approach. By applying DPX approach, designers have made some innovative products. They have achieved good results on design works. The main contributions of the practice summarized as follows:

(1)Designing comfortable smart clothing: for infants' health.

Based on "DPX" thinking approach, we design the experiment of smart clothing for infants' health monitoring. In the action research, we provide the new methods to solve the problem, and address both comfort and sensor accuracy of smart clothing development in real-world application. The combination of state-of-the-art machine learning algorithms and simple pertinent features improved measurement accuracy between 10% and 300%, based on various important evaluation measures. This not only constitutes a significant achievement but also encourages further study of the proposed strategy for interdisciplinary practical fashion education.

(2)Fashionable experience for blind children—design research of intelligent glove featuring perception of chromatic color.

Based on "DPX" thinking approach, we design the intelligent glove for blind children. We have broken a natural limitation. We possibly cannot only observe things through our eyes but also perceive things through temperature through the painting process. A chromatic-color element is combined with a temperature element. The temperature is used to perceive chromatic color, the chromatic-color is sufficiently utilized, and better perception and expression of things are realized. Thus, blind persons can be brought out from the dark world and be enabled to perceive the multicolored world just like a normal person.

5.2.2 Suggestions for Further Research

With the rapid development of science and technology, the design has evolved from the tangible and intangible design to the design of the system, and then to the design of the complex adaptive system. This evolution also brings about changes in the designers' responsibilities. Designers need more comprehensive knowledge to become participants in the work system. Designers should know more about technical information. On the one hand, they should break the knowledge boundary between art and science. On the other hand, they should make self cognition on these two knowledge areas. As fashion designers, we should make full use of information science to build a new relationship between human and clothing for the future world. Clothing will become the carrier of technology and become the interface of human interaction with the world. This is also the vision of the "DPX".

The "X" in "DPX" means the knowledge of different fields. With the development of science and technology, X has been developing dynamically. Research content can become more complex. It is suggested that designers should create more innovative, experimental and new fields for smart clothing. Designers should make efforts on thoroughly recording experimental procedures and results and making tacit knowledge clearer. The knowledge creation process should be developed into a model of thinking. The model could become a very useful tool in the process of research and development of smart clothing.

Most of today's design challenges require analytic and synthetic planning skills that can't be developed through the practice of contemporary design professions alone. Professional design practice involves advanced multi-disciplinary knowledge that presupposes interdisciplinary collaboration and a fundamental change in design education. This knowledge isn't simply a higher level of professional education and practice. It is a qualitatively different form of professional practice. It is emerging in response to the demands of the information society and the knowledge economy to which it gives rise.

We already face the challenges of future design education. on the one hand, If these challenges are not evenly distributed, neither are the skills or capacities that design schools need to meet them. The design education we need today is increasingly similar to the requirements of professional education in engineering—or, perhaps better said, it is increasingly similar to the requirements of education in health care and medicine. To succeed, outstanding professional design requires a foundation based on science and on research. To serve human beings, outstanding professional designers must master an art of human engagement based on ethics and on care. Design education must foster such skills and knowledge.

Reference

[1] LYMBERIS A, OLSSON S. Intelligent biomedical clothing for personal health and disease management: state of the art and future vision [J]. Telemedicine journal and e-health, 2003, 9 (4): 379-386.

[2] BROWN T, WYATT J. Design thinking for social innovation [J]. Stanford social innovation review, 2010, 12(1): 29-43.

[3] BROWN T. Design thinking [J]. Harvard business review, 2008, 86(6): 84-92.

[4] CHO G, RATON B. Smart clothing: technology and applications [M]. Boca Raton: CRC Press, Taylor & Francis Group, 2010: 275.

[5] TAI C C, CHANG CHIEN J R, WANG C Y, et al. Infant monitoring system based on ARM embedded platform [J]. Biomedical engineering: applications, basis and communications, 2008, 20(5): 269-275.

[6] BISHOP M, et al. Pattern recognition and machine learning [M]. New York: Springer, 2006: 738.

[7] RASMUSSEN E. Gaussian processes in machine learning [C]// ML 2003. Lecture notes in computer science, vol 3176. Springer, Berlin, Heidelberg. BOUSQUET O, et al. Advanced lectures on machine learning. Berlin: Springer, 2004: 63-71.

[8] DUNNE L E, SMYTH B, CAULFIELD B. Evaluating the impact of garment structure on wearable sensor performance [C]// 11-13 Oct. 2007. Boston, MA, USA. IEEE international symposiun on wearable computers: IEEE, 2007: 123-124.

[9] THORP O. The invention of the first wearable computer [C]// 19-20 Oct. 1998. Pittsburgh, PA, USA. Second international symposium on wearable computers: IEEE, 1998: 4-8.

[10] POST R, ORTH M, RUSSO P, et al. E-broidery: design and fabrication of textile-based computing [J]. IBM systems journal, 2000, 39(3.4): 840-860.

[11] DIEFFENBANCHER F. Fashion thinking [M]. London: AVA Publishing, 2013: 256.

[12] KELLY G. Body temperature variability (Part 1): a review of the history of body temperature and its variability due to site selection, biological rhythms, fitness, and aging [J]. Alternative medicine review, 2006, 11(4): 278-293.

[13] CHO G, RATON B. Smart clothing: technology and applications [M].

BocaRaton CRC Press: Taylor & Francis Group, 2010: 275.

[14] JESSE G. Customer loyalty and the elements of user experience [J]. Design management review, 2010, 17(1): 35.

[15] LORIGA N, TACCINI D, DE ROSSI, et al. Textile sensing interfaces for cardiopulmonary signs monitoring [C]// 17-18 Jan. 2006. Shanghai, China. Engineering in medicine and biology: IEEE, 2006:7349-7352.

[16] MIGUEL H. Fashion buying and merchandising: from mass-market to luxury retail [M]. USA: Create Space, 2015.

[17] ROBINSON L, JOU H, SPADY D W. Accuracy of parents in measuring body temperature with a tympanic thermometer [J]. BMC family practice, 2005, 6(3): 3-8.

[18] MALINS P, STEED J, FAIRBURN S M, et al. Future textile visions: smart textiles for health and wellness [D]. Aberdeen: Robert Gordon University, 2012.

[19] RANTANEN J, ALFTHAN N, IMPIO J, et al. Smart clothing for the arctic environment [C]// 16-17 Oct. 2000. Atlanta, GA, USA. Fourth international symposium on wearable computers: IEEE, 2000: 15-23.

[20] BERZOWSKA J, COELHO M. Kukkia and Vilkas: kinetic electronic garments [C]// 18-21 Oct. 2005. Osaka, Japan. International symposium on wearable computers: IEEE, 2005: 82-85.

[21] SMITH J, MACLEAN K. Communicating emotion through a haptic link: design space and methodology [J]. International journal of human-computer studies, 2007, 65(4): 376-387.

[22] COOSEMANS J, HERMANS B, PUERS R. Integrating wireless ECG monitoring in textiles [J]. Sensors and actuators: physical, 2006 (130-131): 48-53.

[23] TILLOTSON J. Scentsory design: a holistic approach to fashion as a vehicle to deliver emotional well-being [J]. Fashion practice: the journal of design, creative process & the fashion industry, 2009, 1(1): 33-61.

[24] AGUIRIANO J G. New high tech textiles—new benefits for consumers [C]// Cahn #A. 5th World Conference on Detergents: reinventing the industry: opportunities and challenges. Journal of the American Oil Chemists Society, 2003: 130.

[25] KNIGHT J F, BABER C. A tool to assess the comfort of wearable computers [J].

Human factors: the journal of the human factors and ergonomics society, 2005, 47(1): 77-91.

[26] KIRSTEIN T, COTTET D, GRZYB J, TRÖSTER G. Wearable computing systems-electronic textiles [C]// Tao Xiaoming. In wearable electronics and photonics. Amsterdam, Holland: Elsevier, 2005: 177-197.

[27] ULRICH K T, EPPINGER S D. Product design and development [M]. New York: McGraw-Hill, 1995, 384.

[28] ROBINETTE K M, WHITESTONE J J. The need for improved anthropometric methods for the development of helmet systems [J]. Aviation, space, and environmental medicine, 1994, 65(5): 95.

[29] HOLTZBLATT K, BEYER H. Contextual design: evolved [M]. San Rafael, California, USA. The Morgan & Claypool Publishers Series, 2014.

[30] LANGENHOVE V, HERTLEER C. Smart clothing: a new life [J]. International journal of clothing science and technology, 2004, 16(1-2): 63-72.

[31] DUNNE L E, SMYTH B. Psychophysical elements of wearability [C]// Proceedings of the SIGCHI conference on human. San Jose, CA. Factors in computing systems: association for computing machinery. 2007: 299-302.

[32] DUNNE L L, ASHDOWN S, MCDONALD E. Smart systems: wearable integration of intelligent technology [C]// International centre of excellence for wearable, electronics and smart fashion products (ICEWES) Conference. 2002.

[33] DUNNE L E. Optical bend sensing for wearable goniometry: exploring the comfort/accuracy tradeoff [J]. Research journal of textile & apparel, 2010, 14(4): 73-80.

[34] DUNNE L. Smart clothing in practice: key design barriers to commercialization [J]. Fashion practice: the journal of design, creative process & the fashion, 2010, 2(1): 41-66.

[35] DUNNE L E, SMYTH B, CAULFIELD B. Evaluating the impact of garment structure on wearable sensor performance [C]// 11-13 Oct. 2007. Boston, MA, USA. International symposium on wearable computers: IEEE, 2007: 123-124.

[36] JONAS L. The interaction design foundation: the basics of user experience design [J]. Interaction-design, 2015.

[37] NGUYEN M. The most successful wearables [J]. Wearable-technologies. 2016.

[38] MAHIMKAR M. Wearable technology market analysis-size, share, growth, trends and forecasts to 2022 [J]. Mahadevmahimkar, wordpress, 2015.

[39] MANN S. Smart clothing: the shift to wearable computing [J]. Communications of the ACM, 1996, 39 (8): 23-24.

[40] CHAN D. ESTÈVE, FOURNIOLS J Y, et al. Smart wearable systems: current status and future challenges [J]. Artificial intelligence in medicine, 2012, 56(3): 137-156.

[41] CATRYSSE M, PUERS R, HERTLEER C, et al. Matthys. Towards the

integration of textile sensors in a wireless monitoring suit [J]. Sensors and actuators: physical, 2004, 114(2-3): 302-311.

[42] PERRONE M P, COOPER L N. When networks disagree: ensemble methods for hybrid neural network [J]. Brain and Neural Systems. Providence: Brown University, 1992.

[43] HALL M, FRANK E, HOLMES G, et al. The WEKA data mining software: an update [J]. ACM SIGKDD explorations newsletter, 2009, 11(1): 10 – 18.

[44] MERHOLZ P. Peter in Conversation with Don Norman about UX & Innovation [J]. Adaptive Path, 2007.

[45] NAGAI Y. A sense of design: The embedded motives of nature, culture and future[J]. Principia designae: pre-design, design, and post-design. Tokyo: Springer, 2014: 43-59.

[46] NAGAI Y, CANDY L, Edmonds E. Representations of design thinking [J]. A review of Recent Studies, 2003.

[47] NAGAI Y, JUNAIDY D. Meta-contents of design creativity: extraction of the key concepts that form the sense of design [C]// Proceedings of the Third International Conference on Design Creativity (ICDC2014). The Design Society, 2015: 53-61.

[48] HARROP P, HAYWARD J, DAS R, et al. Wearable Technology 2015-2025: Technologies, Markets, Forecasts [R]. IDTechEx report, 2015.

[49] RAI P, KUMAR P S, OH S, et al. M.P. Agarwal. Smart healthcare textile sensor system for unhindered-pervasive health monitoring [C]. Proc. SPIE. Nanosensors, Biosensors, and Info-Tech Sensors and Systems. 2012.

[50] RAI P, KUMAR P S, OH S, et al. Smart healthcare textile sensor system for unhindered-pervasive health monitoring [C]. SPIE. Smart Structures and Materials+ Nondestructive Evaluation and Health Monitoring. 2012.

[51] PSOMAS S. The five competencies of user experience design [C]. UX Matters, 2007.

[52] PICARD R W. Affective computing: challenges [J]. International Journal of Human-Computer Studies, 2003, 59(1-2): 55-64.

[53] SCHAPIRE R E, FREUND Y, BARTLETT P, et al. Boosting the margin: A new explanation for the effectiveness of voting methods [J]. The annals of statistics, 1998, 26(5): 1651-1686.

[54] JHAJHARIA S, et al. Wearable computing and its application [J]. International journal of computer science and information technologies, 2014, 5(4): 5700-5704.

[55] MANN S. Wearable computing: a first step toward personal imaging [J]. Computer, 1997, 30(2): 25-32.

[56] MANN S. An historical account of the "WearComp" and "WearCam" inventions developed for applications in "Personal Imaging" [C]// 13-14 Oct. 1997.

Cambridge, MA, USA. Digest of Papers, First international symposium on wearable computers: IEEE, 1997: 66-73.

[57] BAHADIR S K, CEBI S, KAHRAMAN C, et al. Developing a smart clothing system for blinds based on information axiom [J]. International journal of computational intelligence systems, 2013, 6(2): 279-292.

[58] TANG S L P, STYLIOS G. An overview of smart technologies for clothing design and engineering [J]. International journal of clothing science and technology, 2005, 18(2): 108-128.

[59] SWALLOW S S, THOMPSON A P. Sensory fabric for ubiquitous interfaces [J]. International journal of human-computer interaction, 2009, 13(2): 147-159.

[60] JANG S, CHO J, JEONG K, et al. Exploring possibilities of ECG electrodes for bio-monitoring smartwear with Cu sputtered fabrics [C]// JACKO J A. Human-computer interaction, interaction platforms and techniques. Berlin, Heidelberg: Springer, 2007: 1130-1137.

[61] FAWCETT T. An introduction to ROC analysis [J]. Pattern recognition letters, 2006, 27(8): 861-874.

[62] VAPNIK V. The nature of statistical learning theory [M]. Berlin: Springer, 2000.

[63] BREIMAN. Random forests [J]. Machine learning, 2001, 45(1): 5-32.

[64] BARFIELD W, MANN S, BAIRD K, et al. Computational clothing and accessories [C]// Fundamentals of Wearable Computers and Augmented Reality. Lawrence Erlbaum Associates, 2001: 471-509.

[65] CHEN CC, SHIH HS. A study of the acceptance of wearable technology for consumers: an analytical network process perspective [J]. International symposium of the analytic hierarchy process, 2014.

[66] TAO X. Smart fibres, fabrics and clothing: fundamentals and applications [M]. Cambridge, England: Woodhead Publishing: Elsevier, 2001.

[67] TAO X. Smart technology for textiles and clothing—introduction and overview [C]// Smart fibers, fabrics and clothing. Cambridge, England: Woodhead Publishing, 2001: 1-6.

[68] TIAN X, TAO X. Mechanical properties of fibre Bragg gratings [C]// Tao X. Smart fibers, fabrics and clothing. Cambridge, England: Woodhead Publishing, 2001: 124-149.

[69] ZHU ZH, LIU T, LI GY, et al. Wearable sensor systems for infants [J]. Sensors, 2015, 15(2): 3721-3749.

Acknowledgements

I have been engaged in the research of innovative design of interdisciplinary integration for ten years. Fortunately enough, I have met many talented persons, who have influenced me and inspired me. The inspirations help me a lot in my research.

First of all, I would like to show my appreciation to Professor Yukari Nagai of Japan Advanced Institute of Science and Technology. Without her consistent and illuminating instruction, this book could not have reached its present form. Professor Nagai, who led me into the creative world of wisdom, I am very lucky to be a student of her.

Secondly, I would like to express my heartfelt gratitude to Associate Professor Takaya Yuizono and Professor Youji Kohda who have instructed and helped me a lot in the past years. I am also greatly indebted to the professors and teachers of the Knowledge Science School of JAIST I benefited a lot from their wonderful courses.

Then, I would like to owe my sincere gratitude to Mr. Kuga and Professor Liu Aijun, who help me get valuable opportunities for studying and working with international experts and professionals. Their enthusiasm and dedication to solving scientific and social problems give me lots of encouragement and confidence for my future design research work. I also owe my sincere gratitude to my friends and my fellow classmates Liu Jing, Shen Tao, Xiao Lin, and Zhao Xiaobo who gave me their help and time in listening to me and helping me work out my problems during the difficult course of the research.

Besides, I would like to give plenty of thanks to Vice-president Professor Ren Wendong for his preface to the series of these books, and plenty of thanks to the Editor Jin Hao of China Textile publishing house for her hard work. As we know, it's not easy to revise the book in English.

Finally, I would like to give special thanks to my family who have encouraged and supported me, many thanks for their loving considerations and great confidence in me all through these years. Love you all.

Ding Wei

July, 2020

内 容 提 要

在设计教育方面，培养跨学科的服装设计人才是新时代教育趋势的重要组成部分，这种趋势是响应新文科建设背景下社会各方对创新和创新思维的绝对需求而产生的。服装设计不仅要回应社会的需求和关切，更要力求为人类生活方式的优化转型做出贡献。本书为读者提供了一种基于多学科交叉融合的服装创新设计新策略即："Design Plus X"（DPX）设计方法，"X"代表了非设计领域的知识集合。设计与跨学科的综合集成将为服装设计师打通学科间的屏障，同时将新的知识框架和愿景带入服装设计教育领域。

本书一方面就知识科学驱动智能服装设计展开深入的设计理论研究；另一方面将服装与电子信息技术、传感器技术、纺织科学等相关领域的前沿技术相结合，展开翔实的智能服装设计实务案例解析。本书可供服装设计专业师生及爱好者参考阅读。

图书在版编目（CIP）数据

跨学科路径下智能服装设计与教育策略研究 / 丁玮著 .-- 北京：中国纺织出版社有限公司，2021.10
（设计学系列成果专著 / 任文东主编）
ISBN 978-7-5180-8356-5

I. ①跨⋯ II. ①丁⋯ III. ① 智能技术—应用—服装设计—教学研究—高等学校 IV. ① TS941.2-39

中国版本图书馆 CIP 数据核字（2021）第 023025 号

责任编辑：金　昊　谢冰雁　　责任校对：楼旭红
责任印制：王艳丽

中国纺织出版社有限公司出版发行
地址：北京市朝阳区百子湾东里 A407 号楼　邮政编码：100124
销售电话：010—67004422　传真：010—87155801
http://www.c-lexlilep.com
中国纺织出版社天猫旗舰店
官方微博 http://weibo.com/2119887771
北京华联印刷有限公司印刷　各地新华书店经销
2021 年 10 月第 1 版第 1 次印刷
开本：787×1092　1/16　印张：10
字数：168 千字　定价：98.00 元